Springer Undergraduate Mathen

Other books in this series

A First Course in Discrete Mathematics *I. Anderson*
Analytic Methods for Partial Differential Equations *G. Evans, J. Blackledge, P. Yardley*
Applied Geometry for Computer Graphics and CAD, Second Edition *D. Marsh*
Basic Linear Algebra, Second Edition *T.S. Blyth and E.F. Robertson*
Basic Stochastic Processes *Z. Brzeźniak and T. Zastawniak*
Complex Analysis *J.M. Howie*
Elementary Differential Geometry *A. Pressley*
Elementary Number Theory *G.A. Jones and J.M. Jones*
Elements of Abstract Analysis *M. Ó Searcóid*
Elements of Logic via Numbers and Sets *D.L. Johnson*
Essential Mathematical Biology *N.F. Britton*
Fields, Flows and Waves: An Introduction to Continuum Models *D.F. Parker*
Further Linear Algebra *T.S. Blyth and E.F. Robertson*
Geometry *R. Fenn*
Groups, Rings and Fields *D.A.R. Wallace*
Hyperbolic Geometry *J.W. Anderson*
Information and Coding Theory *G.A. Jones and J.M. Jones*
Introduction to Laplace Transforms and Fourier Series *P.P.G. Dyke*
Introduction to Ring Theory *P.M. Cohn*
Introductory Mathematics: Algebra and Analysis *G. Smith*
Linear Functional Analysis *B.P. Rynne and M.A. Youngson*
Mathematics for Finance: An Introduction to Financial Engineering *M. Capińksi and T. Zastawniak*
Matrix Groups: An Introduction to Lie Group Theory *A. Baker*
Measure, Integral and Probability, Second Edition *M. Capińksi and E. Kopp*
Multivariate Calculus and Geometry, Second Edition *S. Dineen*
Numerical Methods for Partial Differential Equations *G. Evans, J. Blackledge, P.Yardley*
Probability Models *J.Haigh*
Real Analysis *J.M. Howie*
Sets, Logic and Categories *P. Cameron*
Special Relativity *N.M.J. Woodhouse*
Symmetries *D.L. Johnson*
Topics in Group Theory *G. Smith and O. Tabachnikova*
Vector Calculus *P.C. Matthews*

P.C. Matthews

Vector Calculus

With 63 Figures

Springer

Paul C. Matthews, PhD
School of Mathematical Sciences, University of Nottingham, University Park, Nottingham, NG7 2RD, UK

Cover illustration elements reproduced by kind permission of:
Aptech Systems, Inc., Publishers of the GAUSS Mathematical and Statistical System, 23804 S.E. Kent-Kangley Road, Maple Valley, WA 98038, USA. Tel: (206) 432 - 7855 Fax (206) 432 - 7832 email: info@aptech.com URL: www.aptech.com
American Statistical Association: Chance Vol 8 No 1, 1995 article by KS and KW Heiner 'Tree Rings of the Northern Shawangunks' page 32 fig 2
Springer-Verlag: Mathematica in Education and Research Vol 4 Issue 3 1995 article by Roman E Maeder, Beatrice Amrhein and Oliver Gloor 'Illustrated Mathematics: Visualization of Mathematical Objects' page 9 fig 11, originally published as a CD ROM 'Illustrated Mathematics' by TELOS: ISBN 0-387-14222-3, german edition by Birkhauser: ISBN 3-7643-5100-4.
Mathematica in Education and Research Vol 4 Issue 3 1995 article by Richard J Gaylord and Kazume Nishidate 'Traffic Engineering with Cellular Automata' page 35 fig 2. Mathematica in Education and Research Vol 5 Issue 2 1996 article by Michael Trott 'The Implicitization of a Trefoil Knot' page 14.
Mathematica in Education and Research Vol 5 Issue 2 1996 article by Lee de Cola 'Coins, Trees, Bars and Bells: Simulation of the Binomial Process page 19 fig 3. Mathematica in Education and Research Vol 5 Issue 2 1996 article by Richard Gaylord and Kazume Nishidate 'Contagious Spreading' page 33 fig 1. Mathematica in Education and Research Vol 5 Issue 2 1996 article by Joe Buhler and Stan Wagon 'Secrets of the Madelung Constant' page 50 fig 1.

British Library Cataloguing in Publication Data
Matthews, P.C.
Vector calculus. - (Springer undergraduate mathematics series)
1. Vector analysis 2. Calculus of tensors
I. Title
515.6'3
ISBN 3540761802

Library of Congress Cataloging-in-Publication Data
Matthews, P.C. (Paul Charles), 1962-
Vector calculus / P.C. Matthews.
p. cm. -- (Springer undergraduate mathematics series)
Includes index.
ISBN 3-540-76180-2 (pbk. : acid-free paper)
1. Vector analysis. I. Title. II. Series.
QA433.M38 1998 97-41192
515'.63--dc21 CIP

Springer Undergraduate Mathematics Series ISSN 1615-2085
ISBN 3-540-76180-2 Springer-Verlag London Berlin Heidelberg
Springer Science+Business Media
springeronline.com

Printed in Great Britain
7th printing 2005

Typesetting: Camera ready by author
Printed and bound at the Athenæum Press Ltd., Gateshead, Tyne & Wear
12/3830-6 Printed on acid-free paper SPIN 11338062

Preface

Vector calculus is the fundamental language of mathematical physics. It provides a way to describe physical quantities in three-dimensional space and the way in which these quantities vary. Many topics in the physical sciences can be analysed mathematically using the techniques of vector calculus. These topics include fluid dynamics, solid mechanics and electromagnetism, all of which involve a description of vector and scalar quantities in three dimensions.

This book assumes no previous knowledge of vectors. However, it is assumed that the reader has a knowledge of basic calculus, including differentiation, integration and partial differentiation. Some knowledge of linear algebra is also required, particularly the concepts of matrices and determinants.

The book is designed to be self-contained, so that it is suitable for a programme of individual study. Each of the eight chapters introduces a new topic, and to facilitate understanding of the material, frequent reference is made to physical applications. The physical nature of the subject is clarified with over sixty diagrams, which provide an important aid to the comprehension of the new concepts. Following the introduction of each new topic, worked examples are provided. It is essential that these are studied carefully, so that a full understanding is developed before moving ahead. Like much of mathematics, each section of the book is built on the foundations laid in the earlier sections and chapters. In addition to the worked examples, a section of exercises is included at the middle and at the end of each chapter. Solutions to all the exercises are given at the back of the book, but the student is encouraged to attempt all of the exercises before looking up the answers! At the end of each chapter, a one-page summary is given, listing the most essential points of the chapter.

The first chapter covers the basic concepts of vectors and scalars, the ways in which vectors can be multiplied together and some of the applications of vectors to physics and geometry.

Chapter 2 defines the ways in which vector and scalar quantities can be integrated, covering line integrals, surface integrals and volume integrals. Again, these are illustrated with physical applications.

Techniques for differentiating vectors and scalars are given in Chapter 3, which forms the essential core of the subject of vector calculus. The key concepts of gradient, divergence and curl are defined, which provide the basis for the following chapters.

Chapter 4 introduces a new and powerful notation, suffix notation, for manipulating complicated vector expressions. Quantities that run to several lines using conventional vector notation can be written extremely compactly using suffix notation. One of the main reasons for writing this book is that there are very few other books that make full use of suffix notation, although it is commonly used in undergraduate mathematics courses.

Two important theorems, the divergence theorem and Stokes's theorem, are covered in Chapter 5. These help to tie the subject together, by providing links between the different forms of integrals from Chapter 2 and the derivatives of vectors from Chapter 3.

Chapter 6 covers the general theory of orthogonal curvilinear coordinate systems and describes the two most important examples, cylindrical polar coordinates and spherical polar coordinates.

Chapter 7 introduces a more rigorous, mathematical definition of vectors and scalars, which is based on the way in which they transform when the coordinate system is rotated. This definition is extended to a more general class of objects known as tensors. Some physical examples of tensors are given to aid the understanding of what can be a difficult concept to grasp.

The final chapter gives a brief overview of some of the applications of the subject, including the flow of heat within a body, the mechanics of solids and fluids and electromagnetism.

Table of Contents

1
Vector Algebra

1.1 Vectors and scalars

This book is concerned with the mathematical description of physical quantities. These physical quantities include vectors and scalars, which are defined below.

1.1.1 Definition of a vector and a scalar

A *vector* is a physical quantity which has both magnitude and direction. There are many examples of such quantities, including velocity, force and electric field. A *scalar* is a physical quantity which has magnitude only. Examples of scalars include mass, temperature and pressure.

In this book, vectors will be written in bold italic type (for example, $\boldsymbol{u}$ is a vector) while scalar quantities will be written in plain italic type (for example, a is a scalar). There are two other commonly used ways of denoting vectors which are more convenient when writing by hand: an arrow over the symbol ($\overrightarrow{u}$) or a line under the symbol ($\underline{u}$).

Vectors can be represented diagrammatically by a line with an arrow at the end, as shown in Figure 1.1. The length of the line shows the magnitude of the vector and the arrow indicates its direction. If the vector has magnitude one, it is said to be a *unit vector*. Two vectors are said to be equal if they have the same magnitude and the same direction.

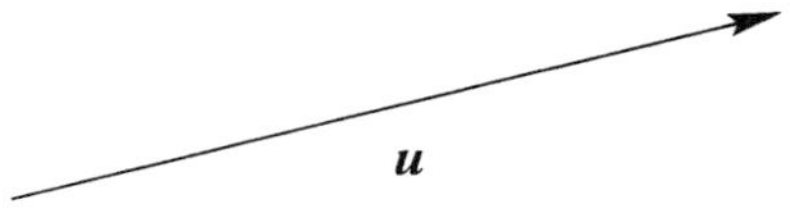

Fig. 1.1. Representation of a vector.

Example 1.1

Classify the following quantities according to whether they are vectors or scalars: energy, electric charge, electric current.

Energy and electric charge are scalars since there is no direction associated with them. Electric current is a vector because it flows in a particular direction.

1.1.2 Addition of vectors

Two vector quantities can be added together by the 'triangle rule' as shown in Figure 1.2. The vector $\boldsymbol{a} + \boldsymbol{b}$ is obtained by drawing the vector $\boldsymbol{a}$ and then drawing the vector $\boldsymbol{b}$ starting from the arrow at the end of $\boldsymbol{a}$.

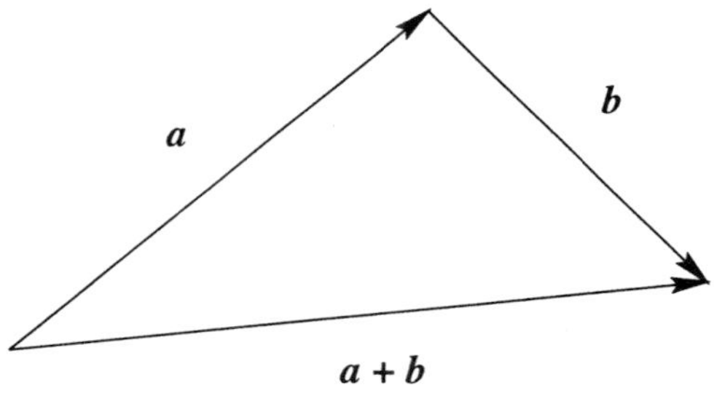

Fig. 1.2. Addition of vectors.

The vector $-\boldsymbol{a}$ is defined as the vector with magnitude equal to that of $\boldsymbol{a}$ but pointing in the opposite direction.

By adding $\boldsymbol{a}$ and $-\boldsymbol{a}$ we obtain the *zero vector*, $\mathbf{0}$. This has magnitude zero and so does not have a direction; nevertheless it is sensible to regard $\mathbf{0}$ as a vector.

1.1.3 Components of a vector

Vectors are often written using a Cartesian coordinate system with axes x, y, z. Such a system is usually assumed to be *right-handed*, which means that a screw rotated from the x-axis to the y-axis would move in the direction of the z-axis. Alternatively, if the thumb of the right hand points in the x direction and the first finger in the y direction, then the second finger points in the z direction.

Suppose that a vector $\boldsymbol{a}$ is drawn in a Cartesian coordinate system and extends from the point (x_1, y_1, z_1) to the point (x_2, y_2, z_2), as shown in Figure 1.3. Then the *components* of the vector are defined to be the three numbers $a_1 = x_2 - x_1$, $a_2 = y_2 - y_1$ and $a_3 = z_2 - z_1$. The vector can then be written in the form $\boldsymbol{a} = (a_1, a_2, a_3)$.

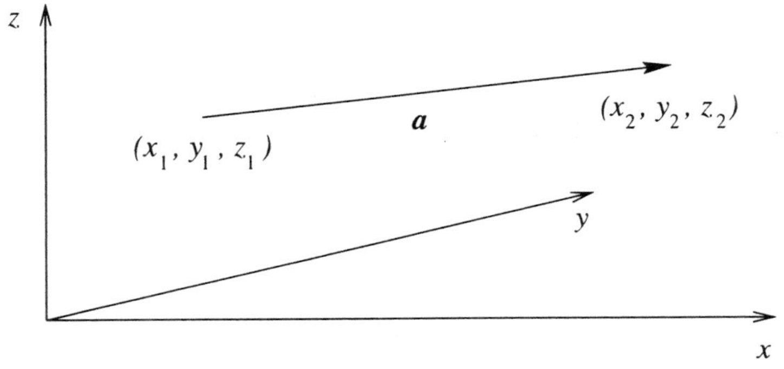

Fig. 1.3. The components of the vector $\boldsymbol{a}$ are $(x_2 - x_1, y_2 - y_1, z_2 - z_1)$.

By introducing three unit vectors $\boldsymbol{e}_1$, $\boldsymbol{e}_2$ and $\boldsymbol{e}_3$, which point along the coordinate axes x, y and z respectively, the vector can also be written in the form $\boldsymbol{a} = a_1\boldsymbol{e}_1 + a_2\boldsymbol{e}_2 + a_3\boldsymbol{e}_3$. Using this form, the sum of the two vectors $\boldsymbol{a}$ and $\boldsymbol{b}$ is $\boldsymbol{a} + \boldsymbol{b} = a_1\boldsymbol{e}_1 + a_2\boldsymbol{e}_2 + a_3\boldsymbol{e}_3 + b_1\boldsymbol{e}_1 + b_2\boldsymbol{e}_2 + b_3\boldsymbol{e}_3 = (a_1 + b_1)\boldsymbol{e}_1 + (a_2 + b_2)\boldsymbol{e}_2 + (a_3 + b_3)\boldsymbol{e}_3$. It follows that vectors can be added simply by adding their components, so that the vector equation $\boldsymbol{c} = \boldsymbol{a} + \boldsymbol{b}$ is equivalent to the three equations $c_1 = a_1 + b_1$, $c_2 = a_2 + b_2$, $c_3 = a_3 + b_3$.

The magnitude of the vector is written $|\boldsymbol{a}|$. It can be deduced from Pythagoras's theorem that the magnitude of the vector can be written in terms of its components as $|\boldsymbol{a}| = \sqrt{a_1^2 + a_2^2 + a_3^2}$.

The position of a point in space (x, y, z) defines a vector which points from the origin of the coordinate system to the point (x, y, z). This vector is called the *position vector* of the point, and is usually denoted by the symbol $\boldsymbol{r}$, with components given by $\boldsymbol{r} = (x, y, z)$.

Example 1.2

The vectors $\boldsymbol{a}$ and $\boldsymbol{b}$ are defined by $\boldsymbol{a} = (1, 1, 1)$, $\boldsymbol{b} = (1, 2, 2)$. Find the magnitudes of $\boldsymbol{a}$ and $\boldsymbol{b}$, and find the vectors $\boldsymbol{a} + \boldsymbol{b}$ and $\boldsymbol{a} - \boldsymbol{b}$.

The magnitude of the vector $\boldsymbol{a}$ is $|\boldsymbol{a}| = \sqrt{1^2 + 1^2 + 1^2} = \sqrt{3}$. The magnitude of $\boldsymbol{b}$ is $|\boldsymbol{b}| = \sqrt{1^2 + 2^2 + 2^2} = 3$. The vector $\boldsymbol{a} + \boldsymbol{b}$ is $(1, 1, 1) + (1, 2, 2) = (2, 3, 3)$ and $\boldsymbol{a} - \boldsymbol{b} = (0, -1, -1)$.

1.2 Dot product

The *dot product* or *scalar product* of two vectors is a scalar quantity. It is written $\boldsymbol{a} \cdot \boldsymbol{b}$ and is defined as the product of the magnitudes of the two vectors and the cosine of the angle between them:

$$\boldsymbol{a} \cdot \boldsymbol{b} = |\boldsymbol{a}||\boldsymbol{b}| \cos\theta. \tag{1.1}$$

A number of properties of the dot product follow from this definition:

- The dot product is commutative, i.e. $\boldsymbol{a} \cdot \boldsymbol{b} = \boldsymbol{b} \cdot \boldsymbol{a}$.
- If the two vectors $\boldsymbol{a}$ and $\boldsymbol{b}$ are perpendicular (orthogonal) then $\boldsymbol{a} \cdot \boldsymbol{b} = 0$.
- Conversely, if $\boldsymbol{a} \cdot \boldsymbol{b} = 0$ then either the two vectors $\boldsymbol{a}$ and $\boldsymbol{b}$ are perpendicular or one of the vectors is the zero vector.
- $\boldsymbol{a} \cdot \boldsymbol{a} = |\boldsymbol{a}|^2$.
- Since the quantity $|\boldsymbol{b}| \cos\theta$ represents the component of the vector $\boldsymbol{b}$ in the direction of the vector $\boldsymbol{a}$, the scalar $\boldsymbol{a} \cdot \boldsymbol{b}$ can be thought of as the magnitude of $\boldsymbol{a}$ multiplied by the component of $\boldsymbol{b}$ in the direction of $\boldsymbol{a}$ (see Figure 1.4).
- The dot product is distributive over addition, i.e. $\boldsymbol{a} \cdot (\boldsymbol{b} + \boldsymbol{c}) = \boldsymbol{a} \cdot \boldsymbol{b} + \boldsymbol{a} \cdot \boldsymbol{c}$. This follows geometrically from the fact that the component of $\boldsymbol{b} + \boldsymbol{c}$ in the direction of $\boldsymbol{a}$ is the same as the component of $\boldsymbol{b}$ in the direction of $\boldsymbol{a}$ plus the component of $\boldsymbol{c}$ in the direction of $\boldsymbol{a}$ (see Figure 1.5).

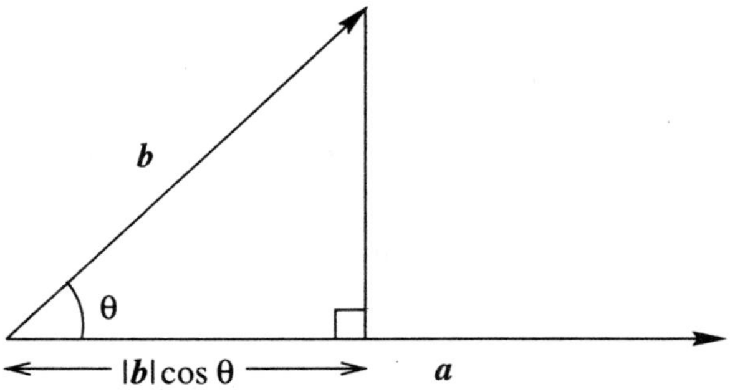

Fig. 1.4. The component of $\boldsymbol{b}$ in the direction of $\boldsymbol{a}$ is $|\boldsymbol{b}| \cos \theta$.

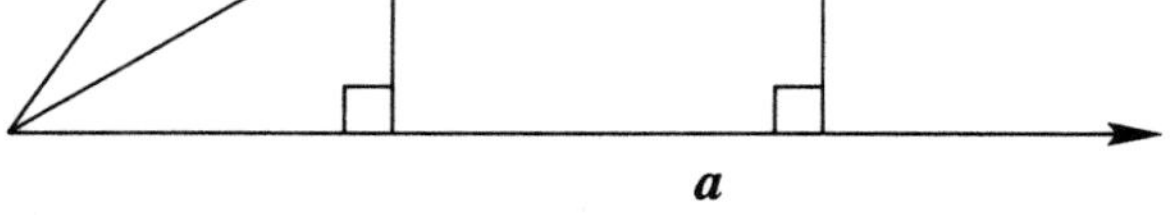

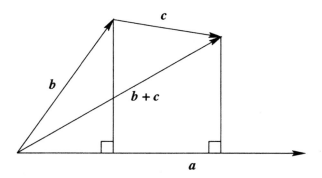

Fig. 1.5. Geometrical demonstration that the dot product is distributive over addition.

A formula for the dot product $\boldsymbol{a} \cdot \boldsymbol{b}$ in terms of the components of the two vectors $\boldsymbol{a}$ and $\boldsymbol{b}$ can be derived from the above properties. Considering first the unit vectors $\boldsymbol{e}_1$, $\boldsymbol{e}_2$ and $\boldsymbol{e}_3$, it follows from the fact that these vectors have magnitude 1 and are orthogonal to each other that

$$\boldsymbol{e}_1 \cdot \boldsymbol{e}_1 = 1,\ \boldsymbol{e}_2 \cdot \boldsymbol{e}_2 = 1,\ \boldsymbol{e}_3 \cdot \boldsymbol{e}_3 = 1,\ \boldsymbol{e}_1 \cdot \boldsymbol{e}_2 = 0,\ \boldsymbol{e}_2 \cdot \boldsymbol{e}_3 = 0,\ \boldsymbol{e}_3 \cdot \boldsymbol{e}_1 = 0.$$

The dot product of $\boldsymbol{a}$ and $\boldsymbol{b}$ is therefore

$$\begin{aligned} \boldsymbol{a} \cdot \boldsymbol{b} &= (a_1\boldsymbol{e}_1 + a_2\boldsymbol{e}_2 + a_3\boldsymbol{e}_3) \cdot (b_1\boldsymbol{e}_1 + b_2\boldsymbol{e}_2 + b_3\boldsymbol{e}_3) \\ &= a_1b_1\boldsymbol{e}_1 \cdot \boldsymbol{e}_1 + a_2b_2\boldsymbol{e}_2 \cdot \boldsymbol{e}_2 + a_3b_3\boldsymbol{e}_3 \cdot \boldsymbol{e}_3 \\ &= a_1b_1 + a_2b_2 + a_3b_3. \end{aligned} \tag{1.2}$$

Example 1.3

Find the dot product of the vectors $(1, 1, 2)$ and $(2, 3, 2)$.

$(1, 1, 2) \cdot (2, 3, 2) = 1 \times 2 + 1 \times 3 + 2 \times 2 = 9.$

Example 1.4

For what value of c are the vectors $(c, 1, 1)$ and $(-1, 2, 0)$ perpendicular?

They are perpendicular when their dot product is zero. The dot product is $-c + 2 + 0$ so the vectors are perpendicular if $c = 2$.

Example 1.5

Show that a triangle inscribed in a circle is right-angled if one of the sides of the triangle is a diameter of the circle.

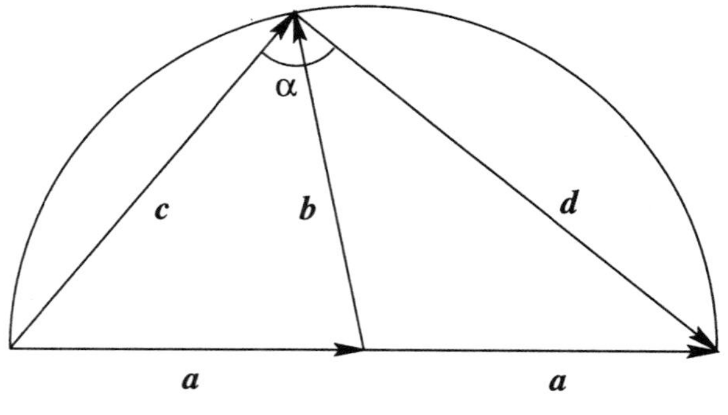

Fig. 1.6. Geometrical construction to show that α is a right angle.

Introduce two vectors $\boldsymbol{a}$ and $\boldsymbol{b}$ as shown in Figure 1.6. Since these two vectors are both along radii of the circle they are of equal magnitude. The two sides $\boldsymbol{c}$ and $\boldsymbol{d}$ of the triangle are then given by $\boldsymbol{c} = \boldsymbol{a}+\boldsymbol{b}$ and $\boldsymbol{d} = \boldsymbol{a}-\boldsymbol{b}$. The dot product of these two vectors is $\boldsymbol{c}\cdot\boldsymbol{d} = (\boldsymbol{a}+\boldsymbol{b})\cdot(\boldsymbol{a}-\boldsymbol{b}) = |\boldsymbol{a}|^2-\boldsymbol{a}\cdot\boldsymbol{b}+\boldsymbol{b}\cdot\boldsymbol{a}-|\boldsymbol{b}|^2 = 0$. Since the dot product is zero the vectors are perpendicular, so the angle α is a right angle. This is just one of many geometrical results that can be obtained using vector methods.

1.2.1 Applications of the dot product

Work done against a force

Suppose that a constant force $\boldsymbol{F}$ acts on a body and that the body is moved a distance $\boldsymbol{d}$. Then the work done against the force is given by the magnitude of the force times the distance moved in the direction opposite to the force; this is simply $-\boldsymbol{F}\cdot\boldsymbol{d}$ (Figure 1.7).

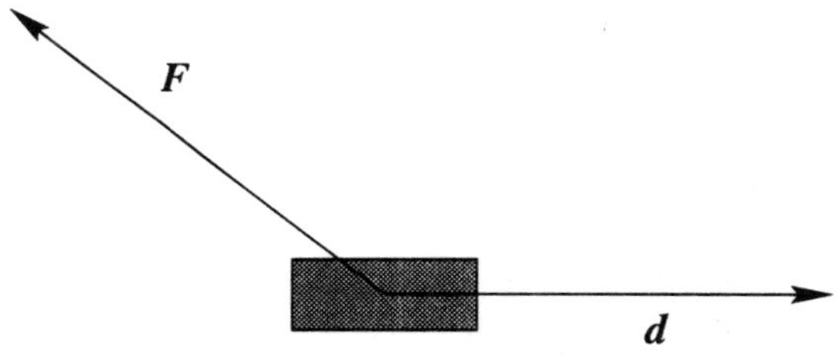

Fig. 1.7. The work done against a force $\boldsymbol{F}$ when an object is moved a distance $\boldsymbol{d}$ is $-\boldsymbol{F}\cdot\boldsymbol{d}$.

Equation of a plane

Consider a two-dimensional plane in three-dimensional space (Figure 1.8). Let $\boldsymbol{r}$ be the position vector of any point in the plane, and let $\boldsymbol{a}$ be a vector perpendicular to the plane. The condition for a point with position vector $\boldsymbol{r}$ to lie in the plane is that the component of $\boldsymbol{r}$ in the direction of $\boldsymbol{a}$ is equal to the perpendicular distance p from the origin to the plane. The general form of the equation of a plane is therefore

$$\boldsymbol{r}\cdot\boldsymbol{a} = \text{constant}.$$

An alternative way to write this is in terms of components. Writing $\boldsymbol{r} = (x, y, z)$ and $\boldsymbol{a} = (a_1, a_2, a_3)$, the equation of a plane becomes

$$a_1x + a_2y + a_3z = \text{constant}. \tag{1.3}$$

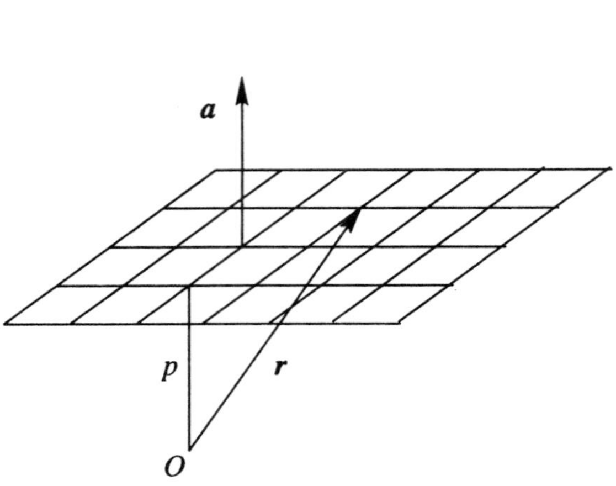

Fig. 1.8. The equation of a plane is $\boldsymbol{r} \cdot \boldsymbol{a} =$ constant.

EXERCISES

1.1 Classify the following quantities according to whether they are vectors or scalars: density, magnetic field strength, power, momentum, angular momentum, acceleration.

1.2 If $\boldsymbol{a} = (2, 0, 3)$ and $\boldsymbol{b} = (1, 0, -1)$, find $|\boldsymbol{a}|$, $|\boldsymbol{b}|$, $\boldsymbol{a} + \boldsymbol{b}$, $\boldsymbol{a} - \boldsymbol{b}$ and $\boldsymbol{a} \cdot \boldsymbol{b}$. What is the angle between the vectors $\boldsymbol{a}$ and $\boldsymbol{b}$?

1.3 If $\boldsymbol{u} = (1, 2, 2)$ and $\boldsymbol{v} = (-6, 2, 3)$, find the component of $\boldsymbol{u}$ in the direction of $\boldsymbol{v}$ and the component of $\boldsymbol{v}$ in the direction of $\boldsymbol{u}$.

1.4 Find the equation of the plane that is perpendicular to the vector $(1, 1, -1)$ and passes through the point $x = 1$, $y = 2$, $z = 1$.

1.5 Use vector methods to show that the diagonals of a rhombus are perpendicular.

1.6 What is the angle between any two diagonals of a cube?

1.7 Use vectors to show that for any triangle, the three lines drawn from each vertex to the midpoint of the opposite side all pass through the same point.

1.3 Cross product

The *cross product* or *vector product* of two vectors is a vector quantity, written $\boldsymbol{a} \times \boldsymbol{b}$. Since it is a vector, its definition must specify both its magnitude and direction. The magnitude of $\boldsymbol{a}\times\boldsymbol{b}$ is $|\boldsymbol{a}||\boldsymbol{b}|\sin\theta$, where θ is the angle between the two vectors $\boldsymbol{a}$ and $\boldsymbol{b}$. The direction of $\boldsymbol{a}\times\boldsymbol{b}$ is perpendicular to both $\boldsymbol{a}$ and $\boldsymbol{b}$ in a right-handed sense, i.e. a right-handed screw rotated from $\boldsymbol{a}$ towards $\boldsymbol{b}$ moves in the direction of $\boldsymbol{a}\times\boldsymbol{b}$ (Figure 1.9). We may therefore write $\boldsymbol{a}\times\boldsymbol{b} = |\boldsymbol{a}||\boldsymbol{b}|\sin\theta\, \boldsymbol{u}$, where $\boldsymbol{u}$ is a unit vector perpendicular to $\boldsymbol{a}$ and $\boldsymbol{b}$ in a right-handed sense.

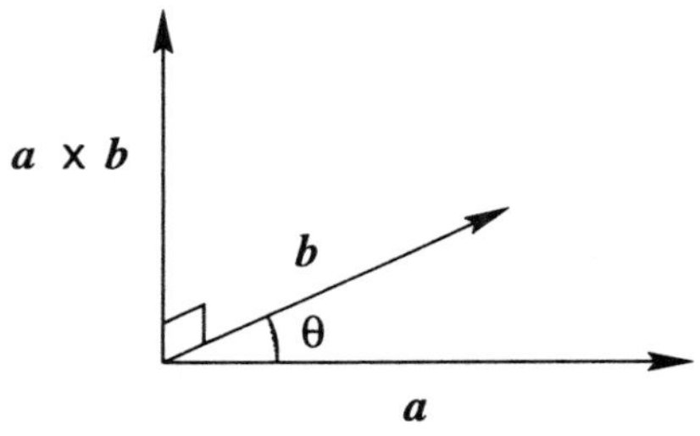

Fig. 1.9. The cross product of $\boldsymbol{a}$ and $\boldsymbol{b}$ is perpendicular to $\boldsymbol{a}$ and $\boldsymbol{b}$, in a right-handed sense.

The cross product has the following properties:

- The cross product is *not* commutative. Because of the right-hand rule, $\boldsymbol{a} \times \boldsymbol{b}$ and $\boldsymbol{b} \times \boldsymbol{a}$ point in opposite directions, so $\boldsymbol{a} \times \boldsymbol{b} = -\boldsymbol{b} \times \boldsymbol{a}$.
- If the two vectors $\boldsymbol{a}$ and $\boldsymbol{b}$ are parallel then $\boldsymbol{a} \times \boldsymbol{b} = \boldsymbol{0}$.
- $\boldsymbol{a} \times \boldsymbol{a} = \boldsymbol{0}$.
- The magnitude of the cross product of $\boldsymbol{a}$ and $\boldsymbol{b}$ is the area of the parallelogram made by the two vectors $\boldsymbol{a}$ and $\boldsymbol{b}$ (Figure 1.10). Similarly the area of the triangle made by $\boldsymbol{a}$ and $\boldsymbol{b}$ is $|\boldsymbol{a} \times \boldsymbol{b}|/2$.
- The cross product of $\boldsymbol{a}$ and $\boldsymbol{b}$ only depends on the component of $\boldsymbol{b}$ perpendicular to $\boldsymbol{a}$. This is apparent from Figure 1.10 since the component of $\boldsymbol{b}$ perpendicular to $\boldsymbol{a}$ is $|\boldsymbol{b}|\sin\theta$.
- The cross product is distributive over addition, i.e. $\boldsymbol{a}\times(\boldsymbol{b}+\boldsymbol{c}) = \boldsymbol{a}\times\boldsymbol{b}+\boldsymbol{a}\times\boldsymbol{c}$. This is demonstrated geometrically in Figure 1.11, where the vector $\boldsymbol{a}$ points into the page. The vectors $\boldsymbol{b}$, $\boldsymbol{c}$ and $\boldsymbol{b} + \boldsymbol{c}$ do not necessarily lie in the page, but from the previous point the cross products of these vectors with $\boldsymbol{a}$ only depend on their projections onto the page. The effect of taking the cross

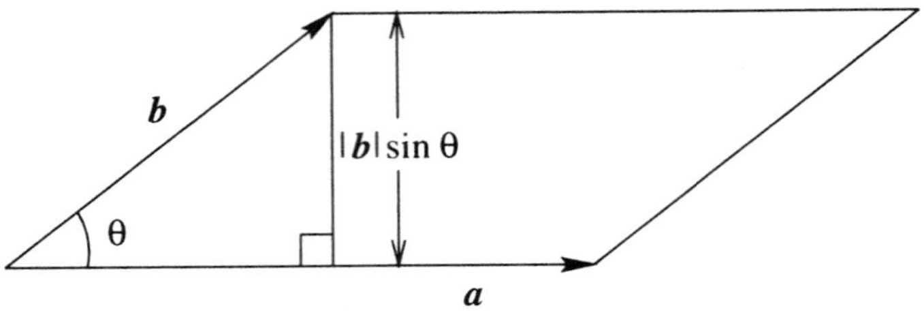

Fig. 1.10. The area of the parallelogram is the length of its base, $|\boldsymbol{a}|$, multiplied by its height, $|\boldsymbol{b}|\sin\theta$.

product with $\boldsymbol{a}$ on any vector is to project it onto the page, rotate through $\pi/2$ clockwise and then multiply by $|\boldsymbol{a}|$. Thus the triangle made by the vectors $\boldsymbol{b}$, $\boldsymbol{c}$ and $\boldsymbol{b}+\boldsymbol{c}$ becomes rotated and scaled as in Figure 1.11 but remains a triangle.

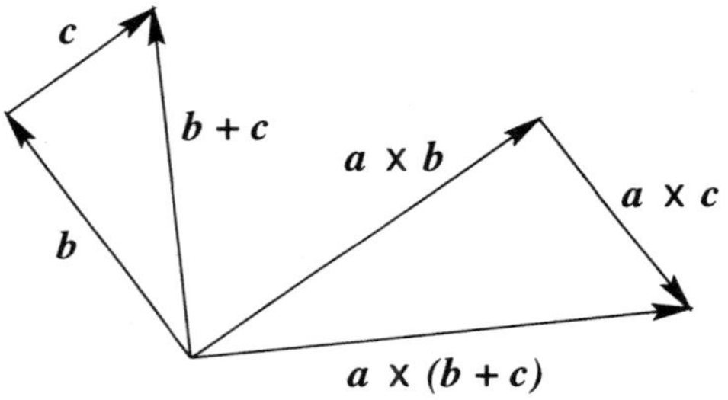

Fig. 1.11. Geometrical demonstration that the cross product is distributive over addition. The vector $\boldsymbol{a}$ points into the page.

A formula for the cross product $\boldsymbol{a}\times\boldsymbol{b}$ in terms of the components of the two vectors $\boldsymbol{a}$ and $\boldsymbol{b}$ can be derived in a similar manner to that carried out for the dot product. Consider first $\boldsymbol{e}_1\times\boldsymbol{e}_2$. Since these two vectors have magnitude 1 and are perpendicular, $\sin\theta = 1$ and the magnitude of $\boldsymbol{e}_1\times\boldsymbol{e}_2$ is 1. The direction of $\boldsymbol{e}_1\times\boldsymbol{e}_2$ is perpendicular to both $\boldsymbol{e}_1$ and $\boldsymbol{e}_2$ in a right-handed sense, so $\boldsymbol{e}_1\times\boldsymbol{e}_2 = \boldsymbol{e}_3$.

It follows that the unit vectors $\boldsymbol{e}_1$, $\boldsymbol{e}_2$ and $\boldsymbol{e}_3$ obey

$$\boldsymbol{e}_1\times\boldsymbol{e}_1 = \boldsymbol{0},\ \boldsymbol{e}_2\times\boldsymbol{e}_2 = \boldsymbol{0},\ \boldsymbol{e}_3\times\boldsymbol{e}_3 = \boldsymbol{0},\ \boldsymbol{e}_1\times\boldsymbol{e}_2 = \boldsymbol{e}_3,\ \boldsymbol{e}_2\times\boldsymbol{e}_3 = \boldsymbol{e}_1,\ \boldsymbol{e}_3\times\boldsymbol{e}_1 = \boldsymbol{e}_2.$$

The cross product of $\boldsymbol{a}$ and $\boldsymbol{b}$ is therefore

$$\begin{aligned} \boldsymbol{a} \times \boldsymbol{b} &= (a_1\boldsymbol{e}_1 + a_2\boldsymbol{e}_2 + a_3\boldsymbol{e}_3) \times (b_1\boldsymbol{e}_1 + b_2\boldsymbol{e}_2 + b_3\boldsymbol{e}_3) \\ &= a_1b_2\,\boldsymbol{e}_1 \times \boldsymbol{e}_2 + a_1b_3\,\boldsymbol{e}_1 \times \boldsymbol{e}_3 + a_2b_1\,\boldsymbol{e}_2 \times \boldsymbol{e}_1 \\ &\quad + a_2b_3\,\boldsymbol{e}_2 \times \boldsymbol{e}_3 + a_3b_1\,\boldsymbol{e}_3 \times \boldsymbol{e}_1 + a_3b_2\,\boldsymbol{e}_3 \times \boldsymbol{e}_2 \\ &= (a_2b_3 - a_3b_2)\boldsymbol{e}_1 + (a_3b_1 - a_1b_3)\boldsymbol{e}_2 + (a_1b_2 - a_2b_1)\boldsymbol{e}_3. \qquad (1.4) \end{aligned}$$

This can also be written as the determinant of a 3×3 matrix as follows:

$$\boldsymbol{a} \times \boldsymbol{b} = \begin{vmatrix} \boldsymbol{e}_1 & \boldsymbol{e}_2 & \boldsymbol{e}_3 \\ a_1 & a_2 & a_3 \\ b_1 & b_2 & b_3 \end{vmatrix}.$$

Example 1.6

Find the cross product of the vectors $(1, 3, 0)$ and $(2, -1, 1)$.

$(1, 3, 0) \times (2, -1, 1) = (3 - 0, 0 - 1, -1 - 6) = (3, -1, -7)$.

Example 1.7

Find a unit vector which is perpendicular to both $(1, 0, 1)$ and $(0, 1, 1)$.

A perpendicular vector is $(1, 0, 1) \times (0, 1, 1) = (-1, -1, 1)$. To make this a unit vector we must divide by its magnitude, which is $\sqrt{3}$, so the unit vector perpendicular to $(1, 0, 1)$ and $(0, 1, 1)$ is $(-1, -1, 1)/\sqrt{3}$.

Example 1.8

What is the area of the triangle which has its vertices at the points $P = (1, 1, 1)$, $Q = (2, 3, 3)$ and $R = (4, 1, 2)$?

First construct two vectors that make up two sides of the triangle. The vector from P to Q is $\boldsymbol{a} = (1, 2, 2)$ and the vector from P to R is $\boldsymbol{b} = (3, 0, 1)$. The cross product of these vectors is $\boldsymbol{a} \times \boldsymbol{b} = (2, 5, -6)$. The area of the triangle is then $|\boldsymbol{a} \times \boldsymbol{b}|/2 = \sqrt{65}/2 \approx 4.03$.

1.3.1 Applications of the cross product

Solid body rotation

Suppose that a solid body is rotating steadily about an axis. What is the velocity vector of a point within the body?

Consider a body rotating with angular velocity Ω (this means that in a time t the body rotates through an angle Ωt radians). Since there is a rotation axis, a vector $\boldsymbol{\Omega}$ can be defined, with magnitude $|\boldsymbol{\Omega}| = \Omega$ and directed along the rotation axis. Since this vector could point in either direction, the following form of the right-hand rule is used to define the direction of $\boldsymbol{\Omega}$: a screw rotating in the same direction as the body moves in the direction of $\boldsymbol{\Omega}$. Alternatively, if the fingers of the right hand point in the direction of the rotation, the thumb of the right hand points in the direction of $\boldsymbol{\Omega}$. This means that for a body which is rotating to the right, $\boldsymbol{\Omega}$ points upwards (Figure 1.12).

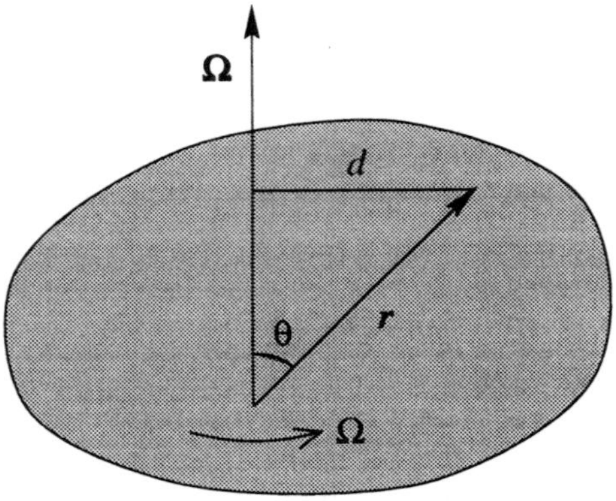

Fig. 1.12. Motion of a rotating body.

Now consider the motion of a point at a position vector $\boldsymbol{r}$, which makes an angle θ with the rotation axis. The speed at which this point moves is Ωd, where d is the perpendicular distance from the point to the rotation axis. Since $d = |\boldsymbol{r}| \sin\theta$ (Figure 1.12), the speed of motion is $v = \Omega|\boldsymbol{r}|\sin\theta$. Note that this is equal to $|\boldsymbol{\Omega} \times \boldsymbol{r}|$. Now consider the direction of the motion. In Figure 1.12, where both $\boldsymbol{\Omega}$ and $\boldsymbol{r}$ lie in the plane of the page, the direction of motion is into the page, perpendicular to both $\boldsymbol{\Omega}$ and $\boldsymbol{r}$ and so in the direction of $\boldsymbol{\Omega} \times \boldsymbol{r}$. Therefore the velocity vector of the point at $\boldsymbol{r}$ is

$$\boldsymbol{v} = \boldsymbol{\Omega} \times \boldsymbol{r}, \tag{1.5}$$

since this vector has both the correct magnitude and the correct direction.

Equation of a straight line

The equation of a straight line can be written in terms of the cross product as follows. Suppose that $\boldsymbol{a}$ is the position vector of a particular fixed point on the line, and that $\boldsymbol{u}$ is a vector pointing along the line (Figure 1.13). Then any point $\boldsymbol{r}$ on the line can be reached from the origin by travelling first along the vector $\boldsymbol{a}$ onto the line and then some multiple of the vector $\boldsymbol{u}$ along the line:

$$\boldsymbol{r} = \boldsymbol{a} + \lambda\boldsymbol{u}, \tag{1.6}$$

where λ is a parameter. This is referred to as the *parametric* form of the equation of a line.

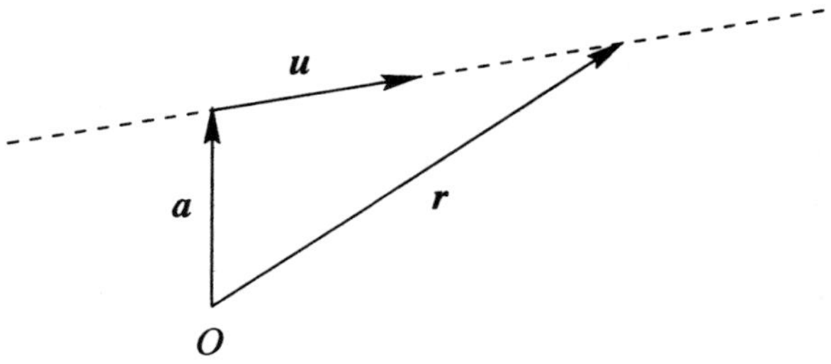

Fig. 1.13. The equation of a line is $\boldsymbol{r} = \boldsymbol{a} + \lambda\boldsymbol{u}$.

To obtain a form of (1.6) that does not involve the parameter λ, the term involving the vector $\boldsymbol{u}$ must be eliminated. This can be done by taking the cross product of (1.6) with $\boldsymbol{u}$. This gives $\boldsymbol{r} \times \boldsymbol{u} = \boldsymbol{a} \times \boldsymbol{u}$. Since the vector $\boldsymbol{a} \times \boldsymbol{u}$ is a constant, it can be relabelled $\boldsymbol{b}$, giving the second form for the equation of a straight line:

$$\boldsymbol{r} \times \boldsymbol{u} = \boldsymbol{b}. \tag{1.7}$$

Physical applications of the cross product

There are many physical quantities that are defined in terms of the cross product. These include the following:

- A particle of mass m has position vector $\boldsymbol{r}$ and is moving with velocity $\boldsymbol{v}$. Its angular momentum about the origin is $\boldsymbol{h} = m\,\boldsymbol{r} \times \boldsymbol{v}$.
- A particle of mass m moves with velocity $\boldsymbol{u}$ in a frame which is rotating with angular velocity $\boldsymbol{\Omega}$. Due to the rotation, the particle experiences a sideways force called the Coriolis force, $\boldsymbol{F} = 2m\,\boldsymbol{u} \times \boldsymbol{\Omega}$. Since the Earth is rotating, this force influences motion on the surface of the Earth. The effect deflects particles to the right in the northern hemisphere and is strongest for motions on large scales such as ocean currents and weather systems.
- A particle with electric charge q moves with velocity $\boldsymbol{v}$ in the presence of a magnetic field $\boldsymbol{B}$. This results in a force, called the Lorentz force, equal to $q\,\boldsymbol{v} \times \boldsymbol{B}$. This is the force which is responsible for the operation of an electric motor.

1.4 Scalar triple product

The *scalar triple product* of three vectors $\boldsymbol{a}$, $\boldsymbol{b}$ and $\boldsymbol{c}$ is defined to be $\boldsymbol{a} \cdot (\boldsymbol{b} \times \boldsymbol{c})$. In fact the brackets here are unnecessary: $(\boldsymbol{a} \cdot \boldsymbol{b}) \times \boldsymbol{c}$ is meaningless since $(\boldsymbol{a} \cdot \boldsymbol{b})$ is a scalar and so cannot be crossed with the vector $\boldsymbol{c}$. Therefore the expression $\boldsymbol{a} \cdot \boldsymbol{b} \times \boldsymbol{c}$ is well defined.

The formula for the scalar triple product in terms of the components of the three vectors $\boldsymbol{a}$, $\boldsymbol{b}$ and $\boldsymbol{c}$ can be obtained using the formula for the cross product (1.4):

$$\boldsymbol{a} \cdot \boldsymbol{b} \times \boldsymbol{c} = a_1 b_2 c_3 - a_1 b_3 c_2 + a_2 b_3 c_1 - a_2 b_1 c_3 + a_3 b_1 c_2 - a_3 b_2 c_1. \tag{1.8}$$

The scalar triple product has a number of properties, listed below. The first four follow directly from (1.8).

- The dot and the cross can be interchanged:

$$\boldsymbol{a} \cdot \boldsymbol{b} \times \boldsymbol{c} = \boldsymbol{a} \times \boldsymbol{b} \cdot \boldsymbol{c}.$$

- The vectors $\boldsymbol{a}$, $\boldsymbol{b}$ and $\boldsymbol{c}$ can be permuted cyclically:

$$\boldsymbol{a} \cdot \boldsymbol{b} \times \boldsymbol{c} = \boldsymbol{b} \cdot \boldsymbol{c} \times \boldsymbol{a} = \boldsymbol{c} \cdot \boldsymbol{a} \times \boldsymbol{b}.$$

- The scalar triple product can be written in the form of a determinant:

$$\boldsymbol{a} \cdot \boldsymbol{b} \times \boldsymbol{c} = \begin{vmatrix} a_1 & a_2 & a_3 \\ b_1 & b_2 & b_3 \\ c_1 & c_2 & c_3 \end{vmatrix}.$$

- If any two of the vectors are equal, the scalar triple product is zero.
- Geometrically, the magnitude of the scalar triple product is the volume of the three-dimensional object known as a parallelepiped formed by the three vectors $\boldsymbol{a}$, $\boldsymbol{b}$ and $\boldsymbol{c}$ (Figure 1.14). This can be shown as follows. The area of the parallelogram forming the base is $|\boldsymbol{b} \times \boldsymbol{c}|$. The height is the vertical component of $\boldsymbol{a}$, which is the magnitude of the component of $\boldsymbol{a}$ in the direction of $\boldsymbol{b} \times \boldsymbol{c}$. This is $|\boldsymbol{a} \cdot \boldsymbol{b} \times \boldsymbol{c}|/|\boldsymbol{b} \times \boldsymbol{c}|$, so the volume is the area of the base multiplied by the height, which is $|\boldsymbol{a} \cdot \boldsymbol{b} \times \boldsymbol{c}|$. Similarly, the volume of the tetrahedron made by the vectors $\boldsymbol{a}$, $\boldsymbol{b}$ and $\boldsymbol{c}$ is $|\boldsymbol{a} \cdot \boldsymbol{b} \times \boldsymbol{c}|/6$.

The scalar triple product of $\boldsymbol{a}$, $\boldsymbol{b}$ and $\boldsymbol{c}$ is often written $[\boldsymbol{a}, \boldsymbol{b}, \boldsymbol{c}]$. This notation highlights the fact that the dot and the cross can be interchanged.

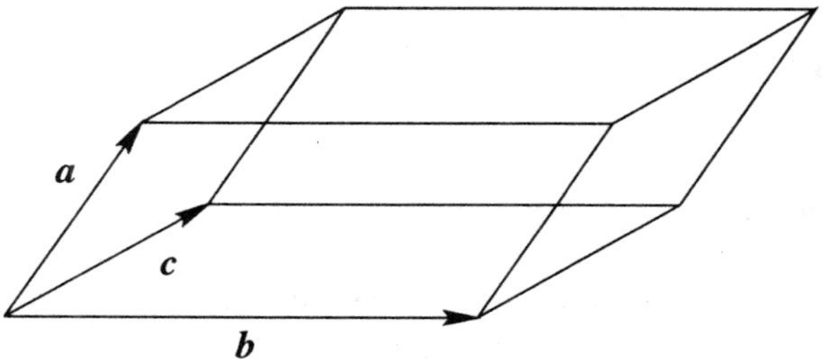

Fig. 1.14. The volume of the object formed by the three vectors $\boldsymbol{a}$, $\boldsymbol{b}$ and $\boldsymbol{c}$ is $|\boldsymbol{a}\cdot\boldsymbol{b}\times\boldsymbol{c}|$.

Example 1.9

Find the scalar triple product of the vectors (1,2,1), (0,1,1) and (2,1,0).

First find the vector $(0,1,1) \times (2,1,0) = (-1,2,-2)$. Now dot this with $(1,2,1)$, giving the answer 1.

Example 1.10

Show that if three vectors lie in a plane, then their scalar triple product is zero.

If $\boldsymbol{a}$, $\boldsymbol{b}$ and $\boldsymbol{c}$ lie in a plane, then the vector $\boldsymbol{b}\times\boldsymbol{c}$ is perpendicular to the plane and hence perpendicular to $\boldsymbol{a}$. Since the dot product of perpendicular vectors is always zero, it follows that $\boldsymbol{a}\cdot\boldsymbol{b}\times\boldsymbol{c} = 0$.

Example 1.11

A particle with mass m and electric charge q moves in a uniform magnetic field $\boldsymbol{B}$. Given that the force $\boldsymbol{F}$ on the particle is $\boldsymbol{F} = q\,\boldsymbol{v}\times\boldsymbol{B}$, where $\boldsymbol{v}$ is the velocity of the particle, show that the particle moves at constant speed.

The equation of motion of the particle is written using Newton's second law, force equals mass times acceleration. The acceleration of the particle is the rate of change of the velocity, written $\dot{\boldsymbol{v}}$, so the equation of motion is

$$q\,\boldsymbol{v}\times\boldsymbol{B} = m\,\dot{\boldsymbol{v}}.$$

Now taking the dot product of both sides of this equation with $\boldsymbol{v}$, the scalar triple product on the left-hand side gives zero since two of the vectors are equal. Hence

$$0 = m\,\dot{\boldsymbol{v}}\cdot\boldsymbol{v} = m\frac{d}{dt}(\boldsymbol{v}\cdot\boldsymbol{v})/2 = m\frac{d}{dt}\left(|\boldsymbol{v}|^2\right)/2,$$

so the speed of the particle, $|\boldsymbol{v}|$, does not change with time.

1.5 Vector triple product

The *vector triple product* of three vectors is $\boldsymbol{a} \times (\boldsymbol{b} \times \boldsymbol{c})$. The brackets are important here, since $\boldsymbol{a} \times (\boldsymbol{b} \times \boldsymbol{c}) \neq (\boldsymbol{a} \times \boldsymbol{b}) \times \boldsymbol{c}$. Since only cross products are involved, the result is a vector. An alternative expression for $\boldsymbol{a} \times (\boldsymbol{b} \times \boldsymbol{c})$ can be obtained by writing out the components. Since

$$\boldsymbol{b} \times \boldsymbol{c} = (b_2c_3 - b_3c_2)\boldsymbol{e}_1 + (b_3c_1 - b_1c_3)\boldsymbol{e}_2 + (b_1c_2 - b_2c_1)\boldsymbol{e}_3,$$

the first component of $\boldsymbol{a} \times (\boldsymbol{b} \times \boldsymbol{c})$ is

$$\begin{aligned} [\boldsymbol{a} \times (\boldsymbol{b} \times \boldsymbol{c})]_1 &= a_2(b_1c_2 - b_2c_1) - a_3(b_3c_1 - b_1c_3) \\ &= b_1(a_2c_2 + a_3c_3) - c_1(a_2b_2 + a_3b_3). \end{aligned}$$

By adding and subtracting the quantity $a_1b_1c_1$, this can be written

$$\begin{aligned} [\boldsymbol{a} \times (\boldsymbol{b} \times \boldsymbol{c})]_1 &= b_1(a_1c_1 + a_2c_2 + a_3c_3) - c_1(a_1b_1 + a_2b_2 + a_3b_3) \\ &= b_1\boldsymbol{a} \cdot \boldsymbol{c} - c_1\boldsymbol{a} \cdot \boldsymbol{b}. \end{aligned}$$

Similar equations hold for the second and third components, so the vector triple product can be expanded as

$$\boldsymbol{a} \times (\boldsymbol{b} \times \boldsymbol{c}) = (\boldsymbol{a} \cdot \boldsymbol{c})\boldsymbol{b} - (\boldsymbol{a} \cdot \boldsymbol{b})\boldsymbol{c}. \tag{1.9}$$

From this result it also follows that

$$(\boldsymbol{a} \times \boldsymbol{b}) \times \boldsymbol{c} = -\boldsymbol{c} \times (\boldsymbol{a} \times \boldsymbol{b}) = -(\boldsymbol{c} \cdot \boldsymbol{b})\boldsymbol{a} + (\boldsymbol{c} \cdot \boldsymbol{a})\boldsymbol{b}. \tag{1.10}$$

Example 1.12

Under what conditions are $\boldsymbol{a} \times (\boldsymbol{b} \times \boldsymbol{c})$ and $(\boldsymbol{a} \times \boldsymbol{b}) \times \boldsymbol{c}$ equal?

By comparing (1.9) with (1.10), the two are equal if $-(\boldsymbol{a} \cdot \boldsymbol{b})\boldsymbol{c} = -(\boldsymbol{c} \cdot \boldsymbol{b})\boldsymbol{a}$. This can alternatively be written $\boldsymbol{b} \times (\boldsymbol{a} \times \boldsymbol{c}) = \boldsymbol{0}$.

Example 1.13

Find an alternative expression for $(\boldsymbol{a} \times \boldsymbol{b}) \cdot (\boldsymbol{c} \times \boldsymbol{d})$.

Since the dot and cross can be interchanged in a scalar triple product,

$$\begin{aligned} (\boldsymbol{a} \times \boldsymbol{b}) \cdot (\boldsymbol{c} \times \boldsymbol{d}) &= \boldsymbol{a} \cdot (\boldsymbol{b} \times (\boldsymbol{c} \times \boldsymbol{d})) \\ &= \boldsymbol{a} \cdot ((\boldsymbol{b} \cdot \boldsymbol{d})\boldsymbol{c} - (\boldsymbol{b} \cdot \boldsymbol{c})\boldsymbol{d}) \\ &= (\boldsymbol{a} \cdot \boldsymbol{c})(\boldsymbol{b} \cdot \boldsymbol{d}) - (\boldsymbol{a} \cdot \boldsymbol{d})(\boldsymbol{b} \cdot \boldsymbol{c}). \end{aligned}$$

1.6 Scalar fields and vector fields

A scalar or vector quantity is said be a *field* if it is a function of position. An example of a scalar field is the temperature inside a room; in general the temperature has a different value at different points in space, so the temperature T is a function of position. This is indicated by writing $T(\boldsymbol{r})$, where $\boldsymbol{r}$ is the position vector of a point in space, $\boldsymbol{r} = (x, y, z)$. Other examples of scalar fields include pressure and density. An example of a vector field is the velocity of the air within a room.

In general, a scalar field T is three-dimensional, i.e. it depends on all three coordinates, $T = T(x, y, z)$. Such fields are difficult to visualise. However, if the scalar field only depends on two coordinates, $T = T(x, y)$, then it can be visualised by sketching a contour plot. To do this, the line $T(x, y) =$ constant is plotted for different values of the constant. For example, consider the scalar field $T(x, y) = x^2 + y^2$. The contour lines are the lines $x^2 + y^2 =$ constant, which are concentric circles centred at the origin, as shown in Figure 1.15(a).

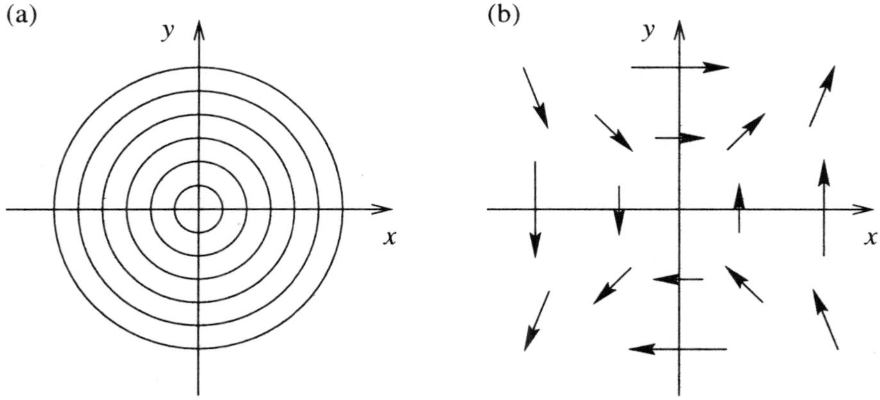

Fig. 1.15. (a) Contours of the scalar field $T(x, y) = x^2 + y^2$. (b) The vector field $\boldsymbol{u}(x, y) = (y, x)$.

Vector fields in two dimensions can also be visualised by a sketch. In this case the simplest procedure is to evaluate the vector field at a sequence of points and draw vectors indicating the magnitude and direction of the vector field at each point. An example of this procedure is the drawing of wind speeds and directions on weather maps. For example, consider the vector field $\boldsymbol{u}(x, y) =$

(y, x). At the point $(1, 0)$, $\boldsymbol{u} = (0, 1)$, so at this point a vector of magnitude 1 pointing in the y direction is drawn. Similarly, at $(0, 1)$, $\boldsymbol{u} = (1, 0)$ and at $(1, 1)$, $\boldsymbol{u} = (1, 1)$. By considering a few additional points, a sketch of the vector field can be built up (Figure 1.15(b)).

Summary of Chapter 1

- A *vector* is a physical quantity with magnitude and direction.
- A *scalar* is a physical quantity with magnitude only.
- In Cartesian coordinates a vector can be written in terms of its *components* as either $\boldsymbol{a} = (a_1, a_2, a_3)$ or $\boldsymbol{a} = a_1\boldsymbol{e}_1 + a_2\boldsymbol{e}_2 + a_3\boldsymbol{e}_3$, where $\boldsymbol{e}_1$, $\boldsymbol{e}_2$ and $\boldsymbol{e}_3$ are unit vectors along the x-, y- and z-axes respectively.
- The *magnitude* of the vector $\boldsymbol{a}$ is $|\boldsymbol{a}| = \sqrt{a_1^2 + a_2^2 + a_3^2}$.
- The *dot product* or *scalar product* of $\boldsymbol{a}$ and $\boldsymbol{b}$ is a scalar,

  $$\boldsymbol{a} \cdot \boldsymbol{b} = |\boldsymbol{a}||\boldsymbol{b}| \cos\theta = a_1 b_1 + a_2 b_2 + a_3 b_3.$$

  This can also be thought of as $|\boldsymbol{a}|$ multiplied by the component of $\boldsymbol{b}$ in the direction of $\boldsymbol{a}$. Applications of the dot product include the work done when moving an object acted on by a force and the equation of a plane.
- The *cross product* or *vector product* of $\boldsymbol{a}$ and $\boldsymbol{b}$ is a vector, $\boldsymbol{a} \times \boldsymbol{b}$, with magnitude $|\boldsymbol{a}||\boldsymbol{b}| \sin\theta$, perpendicular to $\boldsymbol{a}$ and $\boldsymbol{b}$ in a right-handed sense. In component form,

  $$\boldsymbol{a} \times \boldsymbol{b} = (a_2 b_3 - a_3 b_2)\boldsymbol{e}_1 + (a_3 b_1 - a_1 b_3)\boldsymbol{e}_2 + (a_1 b_2 - a_2 b_1)\boldsymbol{e}_3.$$

  The magnitude of $\boldsymbol{a} \times \boldsymbol{b}$ is $|\boldsymbol{a}|$ multiplied by the component of $\boldsymbol{b}$ perpendicular to $\boldsymbol{a}$, which is the area of the parallelogram made by $\boldsymbol{a}$ and $\boldsymbol{b}$. Applications of the cross product include the equation of a straight line and the rotation of a rigid body.
- The *scalar triple product* is $\boldsymbol{a} \cdot \boldsymbol{b} \times \boldsymbol{c} = \boldsymbol{a} \times \boldsymbol{b} \cdot \boldsymbol{c} = \boldsymbol{b} \cdot \boldsymbol{c} \times \boldsymbol{a}$.
- The *vector triple product* is $\boldsymbol{a} \times (\boldsymbol{b} \times \boldsymbol{c}) = (\boldsymbol{a} \cdot \boldsymbol{c})\boldsymbol{b} - (\boldsymbol{a} \cdot \boldsymbol{b})\boldsymbol{c}$.
- A scalar or vector quantity is a *field* if it is a function of position.

EXERCISES

1.8 Find the equation of the straight line which passes through the points $(1,1,1)$ and $(2,3,5)$, (a) in parametric form; (b) in cross product form.

1.9 Using vector methods, prove the sine rule,

$$\frac{\sin A}{a} = \frac{\sin B}{b} = \frac{\sin C}{c} \tag{1.11}$$

and the cosine rule,

$$c^2 = a^2 + b^2 - 2ab\cos C \tag{1.12}$$

for the triangle with angles A, B, C and sides a, b, c in the figure below.

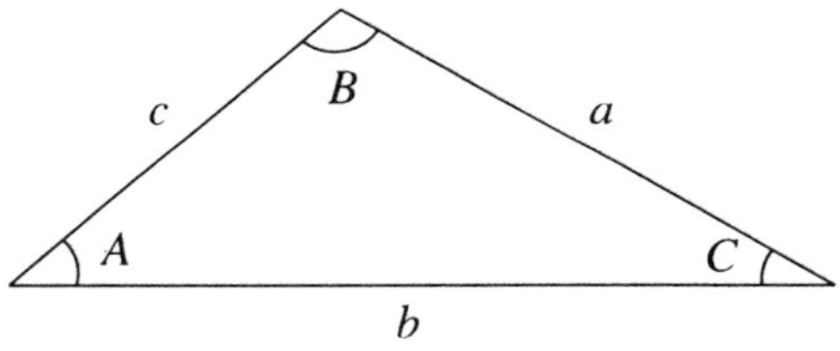

1.10 (a) Show that the set of vectors and the operation of vector addition form a group. (The set of objects $a, b, c, \ldots$ and the operation $\star$ form a group if the following four conditions are satisfied: (i) for any two elements a and b, $a \star b$ is in the set; (ii) $(a \star b) \star c = a \star (b \star c)$; (iii) there is an element I obeying $a \star I = I \star a = a$; (iv) each element a has an inverse a^{-1} such that $a \star a^{-1} = a^{-1} \star a = I$.)
(b) Do the set of vectors and the dot product form a group?
(c) Do the set of vectors and the cross product form a group?

1.11 Simplify the following expressions:
(a) $|\boldsymbol{a} \times \boldsymbol{b}|^2 + (\boldsymbol{a} \cdot \boldsymbol{b})^2$;
(b) $\boldsymbol{a} \times (\boldsymbol{b} \times (\boldsymbol{a} \times \boldsymbol{b}))$;
(c) $(\boldsymbol{a} - \boldsymbol{b}) \cdot (\boldsymbol{b} - \boldsymbol{c}) \times (\boldsymbol{c} - \boldsymbol{a})$;
(d) $(\boldsymbol{a} \times \boldsymbol{b}) \cdot (\boldsymbol{b} \times \boldsymbol{c}) \times (\boldsymbol{c} \times \boldsymbol{a})$.

1.12 The vector $\boldsymbol{x}$ obeys the two equations $\boldsymbol{x} \cdot \boldsymbol{a} = 1$ and $\boldsymbol{x} \times \boldsymbol{a} = \boldsymbol{b}$, where $\boldsymbol{a}$ and $\boldsymbol{b}$ are constant vectors. Solve these equations to find an expression for $\boldsymbol{x}$ in terms of $\boldsymbol{a}$ and $\boldsymbol{b}$. Give a geometrical interpretation of this question.

1.13 Find the equation of the line on which the two planes $\boldsymbol{r} \cdot \boldsymbol{a} = 1$ and $\boldsymbol{r} \cdot \boldsymbol{b} = 1$ meet.

1.14 (a) Express the vector $\boldsymbol{a} \times \boldsymbol{b}$ in the form $\alpha\boldsymbol{a} + \beta\boldsymbol{b} + \gamma\boldsymbol{c}$, assuming that the vectors $\boldsymbol{a}$, $\boldsymbol{b}$ and $\boldsymbol{c}$ are not coplanar.

(b) Hence find an expression for $(\boldsymbol{a} \times \boldsymbol{b} \cdot \boldsymbol{c})^2$ that does not involve any cross products.

(c) Hence find the volume of a tetrahedron made from four equilateral triangles with sides of length 1.

1.15 A particle of mass m at position $\boldsymbol{r}$ and moving with velocity $\boldsymbol{v}$ is subject to a force $\boldsymbol{F}$ directed towards the origin, $\boldsymbol{F} = -f(r)\boldsymbol{r}$. Show that the angular momentum vector $\boldsymbol{h} = m\,\boldsymbol{r} \times \boldsymbol{v}$ is constant.

1.16 Sketch the scalar field $T(x, y) = x^2 - y$.

1.17 Sketch the vector field $\boldsymbol{u}(x, y) = (x + y, -x)$.

2

Line, Surface and Volume Integrals

2.1 Applications and methods of integration

This chapter is concerned with extending the concept of integration to vector quantities and to three dimensions. Before embarking on these more complicated types of integration, however, it is useful to review the concept of integration and some standard techniques for evaluating integrals. It is important that the reader is familiar with these methods, since this will be assumed in the following sections.

2.1.1 Examples of the use of integration

Example 2.1

A rod of length a has a mass per unit length $\rho(x)$ that varies along the length of the rod according to the formula $\rho(x) = 1 + x$. What is the total mass of the rod?

Consider dividing the rod into N small sections, each of length dx_i. The mass of each section is $\rho(x_i)dx_i$. The total mass M of the rod is the sum of the masses of all these sections,

$$M = \sum_{i=1}^{N} \rho(x_i)\, dx_i.$$

The integral of $\rho(x)$ is defined to be the limit of this sum as $N \to \infty$:

$$\int_0^a \rho(x)\, dx = \lim_{N\to\infty} \sum_{i=1}^{N} \rho(x_i) dx_i.$$

The total mass M is therefore

$$M = \int_0^a 1 + x\, dx = \left[x + x^2/2\right]_0^a = a + a^2/2.$$

Example 2.2

A vehicle starts from rest and accelerates uniformly up to a speed of 10 m/s over a time of 20 s. What is the total distance travelled during this time?

The vehicle starts from rest and reaches a speed of 10 m/s after 20 s, so its speed at a time t is $v(t) = t/2$ m/s. In a small time interval dt the distance travelled is $v(t)\, dt = t/2\, dt$. The total distance S travelled in the total time of 20 s is therefore

$$S = \int_0^{20} t/2\, dt = \left[t^2/4\right]_0^{20} = 100\,\text{m}.$$

2.1.2 Integration by substitution

In this method for the evaluation of integrals, a complicated integral is transformed to a simpler one by a substitution or change of variable. In some cases the choice of the change of variable is easy to find, but in others it can be difficult to spot the most sensible substitution. Often, there is more than one possible substitution. Three examples of the application of this method are given below.

Example 2.3

Evaluate $\int x/\sqrt{1-x}\, dx$.

Here, the difficulty is caused by the $\sqrt{1-x}$ in the denominator. This suggests that the appropriate substitution is $u = 1 - x$, so $x = 1 - u$ and $dx = -du$. The integral becomes $\int -(1-u)/\sqrt{u}\, du = \int -1/\sqrt{u} + \sqrt{u}\, du = -2u^{1/2} + 2u^{3/2}/3 + c$, where c is an arbitrary constant of integration. The result can be expressed in terms of the original variable x as $-2\sqrt{1-x}\,(2+x)/3 + c$.

Example 2.4

Evaluate $\int \sqrt{1-x^2}\,dx$.

For integrals involving the quantity $\sqrt{1-x^2}$, the appropriate substitution is $x=\sin\theta$ (or $x=\cos\theta$, which would do equally well). With this choice, $\sqrt{1-x^2}$ becomes $\cos\theta$ and $dx = \cos\theta\,d\theta$. The integral then simplifies to $\int \cos^2\theta\,d\theta$. Integrals of this type, which occur very frequently, are evaluated using the trigonometric formula $\cos^2\theta = (1+\cos 2\theta)/2$, so the value of the integral is $(2\theta+\sin 2\theta)/4+c$. In terms of x this result can be written

$$\int \sqrt{1-x^2}\,dx = \left(\sin^{-1}x + x\sqrt{1-x^2}\right)/2 + c.$$

Example 2.5

Evaluate $\int_0^1 x^2\sqrt{1-x^2}\,dx$.

Again the substitution $x=\sin\theta$ is used. Since the limits of integration are given, these can also be expressed in terms of the new variable θ. When $x=0$, $\theta=0$ and when $x=1$, $\theta=\pi/2$, so the integral becomes $\int_0^{\pi/2}\sin^2\theta\cos^2\theta\,d\theta$. This can be simplified to $\int_0^{\pi/2} 1/4\,\sin^2 2\theta\,d\theta = \int_0^{\pi/2} 1/8\,(1-\cos 4\theta)\,d\theta = \pi/16$.

2.1.3 Integration by parts

Integration by parts is an important and useful technique, used when an integral involves a product of two terms. The integration by parts formula is derived from the product rule for differentiation. Given two functions of x, $u(x)$ and $v(x)$, the rule for the derivative of their product is

$$\frac{d(uv)}{dx} = u\frac{dv}{dx} + v\frac{du}{dx}. \tag{2.1}$$

Integrating this expression and rearranging the terms gives the integration by parts formula:

$$\int u\frac{dv}{dx}\,dx = u\,v - \int v\frac{du}{dx}\,dx. \tag{2.2}$$

As with the case of integration by substitution, some experience is helpful in determining whether this formula will be useful in evaluating an integral, and exactly how to split the integral into the two parts. In general, it is best to choose u to be a function which becomes simpler when differentiated. The following two examples illustrate the use of the method of integration by parts.

Example 2.6

Evaluate $\int x \sin x \, dx$.

In this example we choose $u = x$, $dv/dx = \sin x$, so $v = -\cos x$. Applying the formula (2.2) gives

$$\int x \sin x \, dx = -x \cos x + \int \cos x \, dx = -x \cos x + \sin x + c.$$

Note that it is essential to make the right choice for u and v. If we had chosen $u = \sin x$, $dv/dx = x$ then the resulting integral would have involved $x^2 \cos x$ which is more complicated than the integral we started with.

Example 2.7

Evaluate $\int \exp ax \cos x \, dx$.

For this case two applications of (2.2) are necessary. Choosing $u = \exp ax$, $dv/dx = \cos x$,

$$\int \exp ax \cos x \, dx = \exp ax \sin x - \int a \exp ax \sin x \, dx + c.$$

Now apply (2.2) to the integral on the r.h.s. with $u = a \exp ax$, $dv/dx = \sin x$:

$$\int \exp ax \cos x \, dx = \exp ax \sin x + a \exp ax \cos x - \int a^2 \exp ax \cos x \, dx + c.$$

At this stage it may appear that no progress has been made, since the original integral has reappeared on the right-hand side. However, by rearranging the terms,

$$(1 + a^2) \int \exp ax \cos x \, dx = \exp ax \sin x + a \exp ax \cos x + c$$

and so the value of the original integral is

$$\int \exp ax \cos x \, dx = (\exp ax \sin x + a \exp ax \cos x + c)/(1 + a^2).$$

In this case the choice of u and v does not matter: the result can also be obtained by choosing $u = \cos x$, $dv/dx = \exp ax$, provided that a similar choice, with u chosen to be the trigonometric term and dv/dx chosen to be the exponential term, is made for the second application of (2.2).

2.2 Line integrals

2.2.1 Introductory example: work done against a force

As an introductory example of a line integral, consider a particle moving along a curved path C through space. The particle is acted on by a force $\boldsymbol{F}(\boldsymbol{r})$, which is a vector field. What is the total amount of work done as the particle moves along the curve C?

To answer this question, first divide the curve C into a large number of small pieces. Consider the work done when the particle moves from the position $\boldsymbol{r}$ to $\boldsymbol{r} + \boldsymbol{dr}$ (Figure 2.1). On this small section of the curve C, the work done is $-\boldsymbol{F} \cdot \boldsymbol{dr}$. The total amount of work done W as the particle moves along C is therefore the sum of the contributions from all the small segments of the curve,

$$W = \sum_{i=1}^{N} -\boldsymbol{F}_i \cdot \boldsymbol{dr}_i. \tag{2.3}$$

The *line integral* of $\boldsymbol{F}$ along the curve C is defined by

$$\int_C \boldsymbol{F} \cdot \boldsymbol{dr} = \lim_{N\to\infty} \sum_{i=1}^{N} \boldsymbol{F}_i \cdot \boldsymbol{dr}_i. \tag{2.4}$$

The vector $\boldsymbol{dr}$ is often referred to as a *line element.*

Note that the direction of the integral along the curve C must be specified. If the direction of the curve is reversed, all the line elements $\boldsymbol{dr}$ are reversed and so the value of the integral is multiplied by -1.

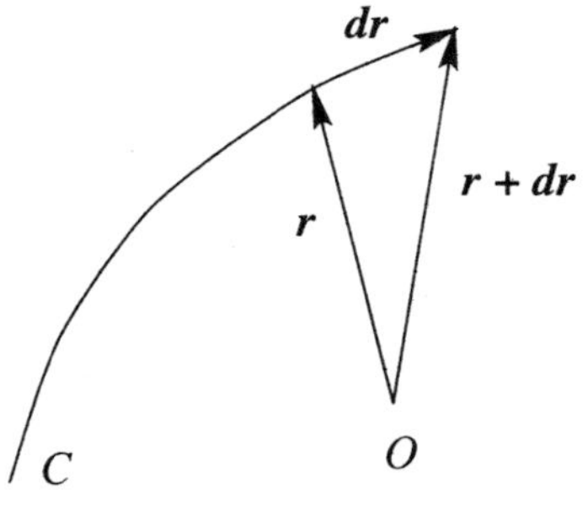

Fig. 2.1. A small section of the curve C is represented by the line element $\boldsymbol{dr}$.

2.2.2 Evaluation of line integrals

Line integrals are evaluated by using a parameter, t for example, together with a formula giving the value of the position vector $\boldsymbol{r}$ in terms of t. This can be regarded as an example of integration by substitution, since

$$\int_C \boldsymbol{F} \cdot \boldsymbol{dr} = \int \boldsymbol{F} \cdot \frac{\boldsymbol{dr}}{dt}\, dt. \tag{2.5}$$

For example, suppose that the curve C is given in terms of t by

$$x = t, \qquad y = t, \qquad z = 2t^2, \tag{2.6}$$

and t lies in the range $0 \leq t \leq 1$. Then as t varies between 0 and 1, the position vector $\boldsymbol{r} = (x, y, z)$ moves along a curve C in space connecting the points $(0, 0, 0)$ and $(1, 1, 2)$. Suppose that the vector field $\boldsymbol{F}$ is given by $\boldsymbol{F} = (y, x, z)$. To evaluate the line integral, both $\boldsymbol{F}$ and $\boldsymbol{dr}/dt$ must be written in terms of t. Substituting (2.6) into the expression for $\boldsymbol{F}$ gives $\boldsymbol{F} = (t, t, 2t^2)$, and

$$\frac{\boldsymbol{dr}}{dt} = \left(\frac{dx}{dt}, \frac{dy}{dt}, \frac{dz}{dt}\right) = (1, 1, 4t).$$

The line integral can now be evaluated:

$$\int_C \boldsymbol{F} \cdot \boldsymbol{dr} = \int_0^1 (t, t, 2t^2) \cdot (1, 1, 4t)\, dt = \int_0^1 2t + 8t^3\, dt = 3.$$

In this first example, the parametric form of the curve C was given. If the curve C is given in a different form, then a parametric form must be constructed so that the line integral can be evaluated. For example, suppose now that $\boldsymbol{F} = (y, x, z)$ as before, but C is the straight line connecting the origin to the point $(1, 2, 3)$. The way in which the curve C is parametrised is not unique, so we can make the arbitrary choice $x = t$. Since x varies between 0 and 1 along the line, this is also the range for t. The end point of C is $(1, 2, 3)$, so y and z must be given by $y = 2t$, $z = 3t$, and so $\boldsymbol{dr} = (1, 2, 3)\, dt$. The value of the integral is therefore

$$\int_C \boldsymbol{F} \cdot \boldsymbol{dr} = \int_0^1 (2t, t, 3t) \cdot (1, 2, 3)\, dt = \int_0^1 13t\, dt = 6.5.$$

Line integrals sometimes occur over curves that are closed, i.e. when the starting point and end point of the curve are equal. In this case the integral is written using the symbol $\oint$, which indicates that the integral is along a closed curve. For example, consider the integral of $\boldsymbol{F} = (y, x, z)$ around the closed curve given by $x = \cos\theta$, $y = \sin\theta$, $z = 0$, where $0 \leq \theta \leq 2\pi$. Here, as θ varies, the curve C describes a circle in the x, y plane. The line element $\boldsymbol{dr}$ is expressed in terms of the parameter θ as $\boldsymbol{dr} = (dx, dy, dz) = (-\sin\theta, \cos\theta, 0)\, d\theta$, so the value of the line integral is

$$\begin{aligned}\oint_C \boldsymbol{F}\cdot\boldsymbol{dr} &= \int_0^{2\pi} (\sin\theta, \cos\theta, 0)\cdot(-\sin\theta, \cos\theta, 0)\, d\theta \\ &= \int_0^{2\pi} -\sin^2\theta + \cos^2\theta\, d\theta \\ &= \int_0^{2\pi} \cos 2\theta\, d\theta = \left[1/2\,\sin 2\theta\right]_0^{2\pi} = 0.\end{aligned}$$

The line integral of a vector field $\boldsymbol{F}$ around a closed curve C is often called the *circulation* of $\boldsymbol{F}$ around C.

Example 2.8

Evaluate the line integral of the vector field $\boldsymbol{u} = (xy, z^2, x)$ along the curve given by $x = 1 + t$, $y = 0$, $z = t^2$, $0 \le t \le 3$.

First write $\boldsymbol{u}$ and $\boldsymbol{dr}$ in terms of t: $\boldsymbol{u} = (0, t^4, 1 + t)$, $\boldsymbol{dr} = (1, 0, 2t)\, dt$. The value of the integral is therefore

$$\int_C \boldsymbol{F}\cdot\boldsymbol{dr} = \int_0^3 (0, t^4, 1+t)\cdot(1, 0, 2t)\, dt = \int_0^3 2t + 2t^2\, dt = \left[t^2 + 2t^3/3\right]_0^3 = 27.$$

Example 2.9

Find the line integral of $\boldsymbol{F} = (y, -x, 0)$ along the curve consisting of the two straight line segments (a) $y = 1$, $0 \le x \le 1$, (b) $x = 1$, $1 \le y \le 2$.

Here, the contributions from the two line segments must be taken separately. On section (a), using x as the parameter, we have

$$\int_0^1 (1, -x, 0)\cdot(dx, 0, 0) = 1.$$

Similarly on section (b) we have

$$\int_1^2 (y, -1, 0)\cdot(0, dy, 0) = -1.$$

Therefore the total value of the integral is 0.

Example 2.10

Find the circulation of the vector $\boldsymbol{F} = (y, -x, 0)$ around the unit circle, $x^2 + y^2 = 1$, $z = 0$, taken in an anticlockwise direction.

The circle is written in terms of a parameter θ as $x = \cos\theta$, $y = \sin\theta$, $z = 0$, $0 \le \theta \le 2\pi$. The value of the integral is

$$\oint_C \boldsymbol{F}\cdot\boldsymbol{dr} = \int_0^{2\pi} (\sin\theta, -\cos\theta, 0)\cdot(-\sin\theta, \cos\theta, 0)\, d\theta = \int_0^{2\pi} -1\, d\theta = -2\pi.$$

2.2.3 Conservative vector fields

A vector field $\boldsymbol{F}$ is said to be *conservative* if it has the property that the line integral of $\boldsymbol{F}$ around any closed curve C is zero:

$$\oint_C \boldsymbol{F} \cdot \boldsymbol{dr} = 0. \tag{2.7}$$

An equivalent definition is that $\boldsymbol{F}$ is conservative if the line integral of $\boldsymbol{F}$ along a curve only depends on the endpoints of the curve, not on the path taken by the curve,

$$\int_{C_1} \boldsymbol{F} \cdot \boldsymbol{dr} = \int_{C_2} \boldsymbol{F} \cdot \boldsymbol{dr} \tag{2.8}$$

where C_1 and C_2 are any two curves that have the same endpoints but different paths (Figure 2.2).

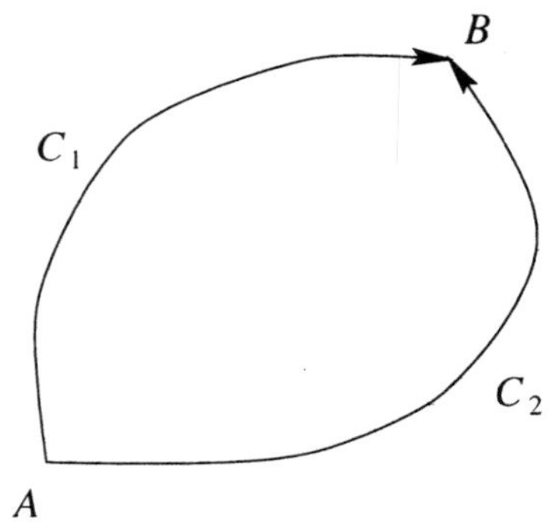

Fig. 2.2. A vector field is conservative if the line integrals along two different curves C_1 and C_2 from A to B are equal.

The equivalence of these two definitions can be demonstrated as follows. Consider two curves C_1 and C_2 that start from the point A and end at the point B (Figure 2.2). Let C be the closed curve that starts from the point A, follows the curve C_1 to the point B and then follows the curve C_2 in the reverse direction to return to A. Then

$$\oint_C \boldsymbol{F} \cdot \boldsymbol{dr} = \int_{C_1} \boldsymbol{F} \cdot \boldsymbol{dr} - \int_{C_2} \boldsymbol{F} \cdot \boldsymbol{dr} \tag{2.9}$$

since the effect of the reversed direction of the integral along C_2 is to change the sign of the integral. From this equation it follows that if the integral around C is zero, then the integrals along C_1 and C_2 are equal, and similarly, if the integrals along C_1 and C_2 are equal, then the integral around C is zero.

Conservative vector fields are of great importance, since many physical examples of vector fields are conservative. Consider for example the Earth's gravitational field, $\boldsymbol{g}$. A particle of mass m experiences a force $m\boldsymbol{g}$, so the work done in moving the particle along a path C from point A to point B is just minus the line integral of $m\boldsymbol{g}$ along C. However, we know physically that the work done only depends on the position of the points A and B – in fact the work done is simply the difference in the potential energy of the particle at A and B. Equivalently, if the particle is moved around but returns to its starting point, the total work done is zero. Therefore, the Earth's gravitational field is an example of a conservative vector field.

Example 2.11

By considering the line integral of $\boldsymbol{F} = (y, x^2 - x, 0)$ around the square in the x, y plane connecting the four points $(0,0)$, $(1,0)$, $(1,1)$ and $(0,1)$, show that $\boldsymbol{F}$ cannot be a conservative vector field.

This line integral consists of four parts (Figure 2.3). On the first section

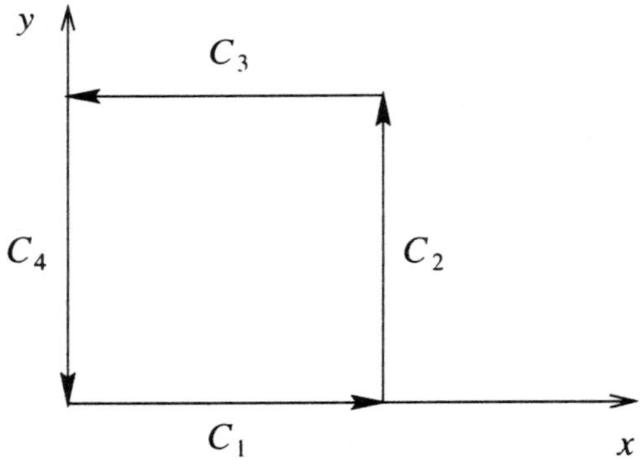

Fig. 2.3. The line integral around the square is split into four straight sections.

C_1, from $(0,0)$ to $(1,0)$, $d\boldsymbol{r} = (dx, 0, 0)$ and $y = 0$ so $\boldsymbol{F} = (0, x^2 - x, 0)$ and $\boldsymbol{F} \cdot d\boldsymbol{r} = 0$. On the second section C_2, $d\boldsymbol{r} = (0, dy, 0)$ and $x = 1$ so $\boldsymbol{F} = (y, 0, 0)$ and again $\boldsymbol{F} \cdot d\boldsymbol{r} = 0$. On the third section C_3, $d\boldsymbol{r} = (dx, 0, 0)$ and $y = 1$ so $\boldsymbol{F} = (1, x^2 - x, 0)$ and the contribution to the line integral is

$$\int_1^0 1\, dx = -1.$$

Finally, on the fourth section C_4 of the square $\boldsymbol{F} \cdot d\boldsymbol{r} = 0$ again, so the total value of the integral is -1. Since the integral around the closed circuit is non-zero, $\boldsymbol{F}$ cannot be conservative.

2.2.4 Other forms of line integrals

The line integral $\int_C \boldsymbol{F} \cdot \boldsymbol{dr}$ is the most important type of line integral, but there are two other forms of the line integral which can occur. These are

$$\int_C \phi \, \boldsymbol{dr} \qquad \text{and} \qquad \int_C \boldsymbol{F} \times \boldsymbol{dr}$$

where ϕ is a scalar field and $\boldsymbol{F}$ is a vector field. Note that in each of these cases the result of the integral is a vector quantity. These integrals can be evaluated using a parameter, as in the following examples.

Example 2.12

Evaluate the line integral

$$\int_C x + y^2 \, \boldsymbol{dr},$$

where C is the parabola $y = x^2$ in the plane $z = 0$ connecting the points $(0, 0, 0)$ and $(1, 1, 0)$.

The curve can be written in terms of a parameter t as $x = t$, $y = t^2$, $z = 0$, $0 \le t \le 1$, so $\boldsymbol{dr} = (1, 2t, 0)\, dt$. The value of the integral is therefore

$$\begin{aligned} \int_C x + y^2 \, \boldsymbol{dr} &= \int_0^1 (t + t^4)(1, 2t, 0)\, dt \\ &= \boldsymbol{e}_1 \left(\int_0^1 t + t^4 \, dt \right) + \boldsymbol{e}_2 \left(\int_0^1 2t^2 + 2t^5 \, dt \right) = 0.7\boldsymbol{e}_1 + \boldsymbol{e}_2. \end{aligned}$$

Example 2.13

Evaluate the line integral

$$\int_C \boldsymbol{F} \times \boldsymbol{dr},$$

where $\boldsymbol{F}$ is the vector field $(y, x, 0)$ and C is the curve $y = \sin x$, $z = 0$, between $x = 0$ and $x = \pi$.

The curve can be written as $x = t$, $y = \sin t$, $z = 0$, $0 \le t \le \pi$. Then $\boldsymbol{F} = (\sin t, t, 0)$ and $\boldsymbol{dr} = (1, \cos t, 0)\, dt$, so $\boldsymbol{F} \times \boldsymbol{dr} = (0, 0, \sin t \cos t - t)\, dt$ and the integral is

$$\int_C \boldsymbol{F} \times \boldsymbol{dr} = \boldsymbol{e}_3 \int_0^\pi \sin t \cos t - t \, dt = 1/2 \left[\sin^2 t - t^2\right]_0^\pi \boldsymbol{e}_3 = -\pi^2/2\, \boldsymbol{e}_3.$$

EXERCISES

2.1 Evaluate the line integral

$$\int_C \boldsymbol{F} \cdot d\boldsymbol{r} \qquad \text{where} \qquad \boldsymbol{F} = (5z^2, 2x, x + 2y) \tag{2.10}$$

and the curve C is given by $x = t$, $y = t^2$, $z = t^2$, $0 \le t \le 1$.

2.2 Evaluate the line integral of the same vector field $\boldsymbol{F}$ given in (2.10) along the straight line joining the points $(0, 0, 0)$ and $(1, 1, 1)$. Is $\boldsymbol{F}$ a conservative vector field?

2.3 Find the line integral of the vector field $\boldsymbol{u} = (y^2, x, z)$ along the curve given by $z = y = \mathrm{e}^x$ from $x = 0$ to $x = 1$.

2.4 Find the line integral $\oint_C \boldsymbol{r} \times d\boldsymbol{r}$ where the curve C is the ellipse $x^2/a^2 + y^2/b^2 = 1$ taken in an anticlockwise direction. What do you notice about the magnitude of the answer?

2.3 Surface integrals

2.3.1 Introductory example: flow through a pipe

Suppose that fluid flows with velocity $\boldsymbol{u}$ through a pipe. What is the total volume of fluid passing through the pipe per unit time (Figure 2.4)? This

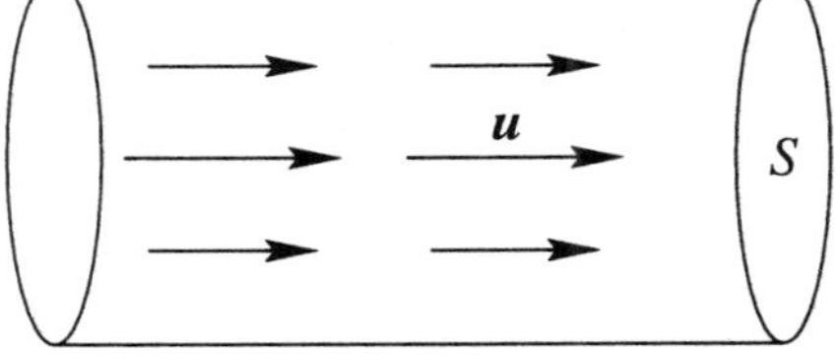

Fig. 2.4. Fluid flows with velocity $\boldsymbol{u}$ along a pipe. The rate at which it crosses the surface S at the end of the pipe is an example of a surface integral.

volume flow rate is often called the *flux* of fluid through the pipe, or the flux of fluid across the surface S that forms the end of the pipe. We will consider this question in several different cases, beginning with the simplest case and progressing to more complicated examples.

Suppose first that the pipe has cross-sectional area A and that the velocity $\boldsymbol{u}$, which in general can be a function of space and time, is a constant and is directed parallel to the walls of the pipe, with speed $|\boldsymbol{u}| = U_0$. In this case, the fluid moves along the pipe as if it were a solid block. In a time t, the fluid moves a distance $U_0 t$, so a 'block' of fluid of volume $U_0 t A$ emerges from the end of the pipe. The flow rate Q, or flux, of fluid through the pipe is therefore this volume divided by the time t, giving $Q = U_0 A$.

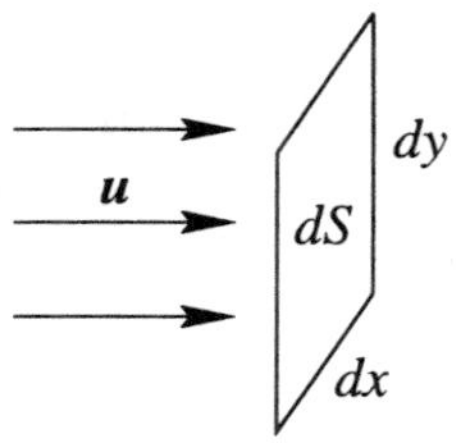

Fig. 2.5. Enlargement of a small surface element dS forming part of the surface S.

Now suppose that the flow is again directed parallel to the walls of the pipe but that the speed of the flow depends on the position within the pipe, so $|\boldsymbol{u}| = U_0(x, y)$, and that the pipe has a square cross-section with the walls at $x = 0, 1$ and $y = 0, 1$. Now consider a small *surface element* with area dS on the surface S, which is a small rectangle with sides of length dx and dy located at the point (x, y) on the surface S, so that $dS = dx\,dy$ (Figure 2.5). Following the argument of the previous paragraph, the flux dQ of fluid across this surface element dS is $dQ = U_0(x, y)\,dS = U_0(x, y)\,dx\,dy$. To calculate the total flux Q across the surface S, we need to add up the contributions from all the small surface elements dS. This sum of contributions becomes an integral, but since the surface is two-dimensional the resulting integral is a surface integral or double integral, representing integration in both the x and y directions:

$$Q = \iint_S U_0(x, y)\,dS = \iint_S U_0(x, y)\,dx\,dy. \tag{2.11}$$

In the above example, the fluid flow direction is perpendicular to the surface S. Consider now the case where the vector field $\boldsymbol{u}$ and the surface S are both arbitrary. In general, S may represent a curved surface. Again we consider a small surface element dS (Figure 2.6) and compute the flux of $\boldsymbol{u}$ across dS. Now if $\boldsymbol{u}$ is perpendicular to dS, this flux is just $|\boldsymbol{u}|dS$, but if $\boldsymbol{u}$ is not perpendicular

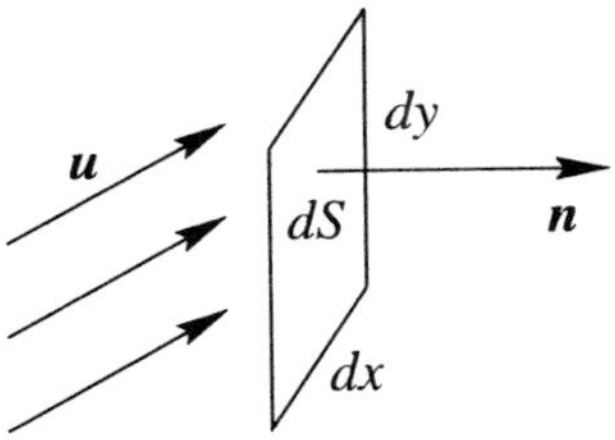

Fig. 2.6. The unit normal vector $\boldsymbol{n}$ is perpendicular to the surface element dS.

to dS, only the component of $\boldsymbol{u}$ perpendicular to dS contributes to the flux across dS. To extract this component it is necessary to introduce a *normal vector* $\boldsymbol{n}$ to the surface dS, with the properties that $\boldsymbol{n}$ is perpendicular to dS and that $\boldsymbol{n}$ is a unit vector, $|\boldsymbol{n}| = 1$. The component of $\boldsymbol{u}$ perpendicular to dS is then the component of $\boldsymbol{u}$ in the direction of $\boldsymbol{n}$, which is just $\boldsymbol{u} \cdot \boldsymbol{n}$ (see Section 1.2), so the flux across the surface element dS is $\boldsymbol{u} \cdot \boldsymbol{n}\, dS$. The total flux across the surface S is given by the surface integral

$$Q = \iint_S \boldsymbol{u} \cdot \boldsymbol{n}\, dS = \iint_S \boldsymbol{u} \cdot \boldsymbol{n}\, dx\, dy.$$

Note that the direction of the normal vector $\boldsymbol{n}$ has not been uniquely specified: the vector $\boldsymbol{n}$ could have been chosen to point in the opposite direction in Figure 2.6, and this would change the sign of the answer. Therefore the direction of $\boldsymbol{n}$ must be specified when a surface integral is written down.

Surface integrals often occur over surfaces which are closed. In this case, the normal to the surface which points outward is used (Figure 2.7). To indicate that the surface is closed, a circle is sometimes drawn through the integral, in a similar manner to the notation used for line integrals around closed curves:

$$\oiint_S \boldsymbol{u} \cdot \boldsymbol{n}\, dS.$$

2.3.2 Evaluation of surface integrals

Surface integrals can be evaluated by carrying out two successive integrations. Consider first the evaluation of (2.11) where S is the square surface given by $0 \le x \le 1$, $0 \le y \le 1$. The surface integral is then

$$Q = \iint_S U_0(x, y)\, dS = \int_0^1 \int_0^1 U_0(x, y)\, dx\, dy.$$

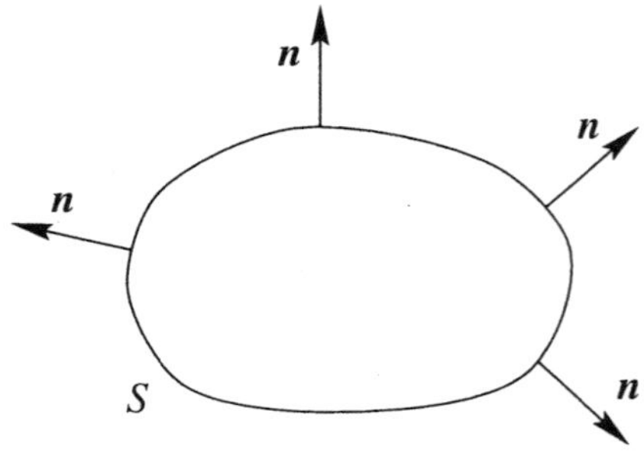

Fig. 2.7. For a closed surface, the convention is that the normal points outward.

There is some potential ambiguity here, since it is not immediately clear which of the integral signs refers to the x integration and which to the y integration. The convention adopted is that the integrals are 'nested', so that the first integral sign represents the y integral and the second one represents the x integral. The double integral is then interpreted as

$$Q = \int_0^1 \left(\int_0^1 U_0(x, y)\, dx \right) dy.$$

It is also assumed that the inner integral, the x integral in the above equation, is to be evaluated first. For example, suppose that $U_0(x, y) = (x - x^2)(y - y^2)$. The inner, x, integral is evaluated first, and within this inner integral y is regarded as a constant. Carrying out the first integral gives

$$\begin{aligned} Q &= \int_0^1 \int_0^1 (x - x^2)(y - y^2)\, dx\, dy \\ &= \int_0^1 \left[x^2/2 - x^3/3\right]_0^1 (y - y^2)\, dy \\ &= \int_0^1 1/6\, (y - y^2)\, dy. \end{aligned}$$

The double integral has now been reduced to a single integral which can be evaluated in the usual way:

$$Q = \int_0^1 1/6\, (y - y^2)\, dy = 1/6\, \left[y^2/2 - y^3/3\right]_0^1 = 1/36.$$

Consider now the case where the surface is circular. In the original example of pipe flow, this corresponds to a cylindrical pipe. If the surface is circular, it is best to use polar coordinates to evaluate the integral. Polar coordinates (r, θ) are related to Cartesian coordinates (x, y) by $x = r\cos\theta$, $y = r\sin\theta$. The angle

θ is measured in radians and ranges from 0 to 2π. The area element dS can be computed by considering a small angle $d\theta$ and a small change in radius dr as shown in Figure 2.8. If dr and $d\theta$ are both small then the corresponding area

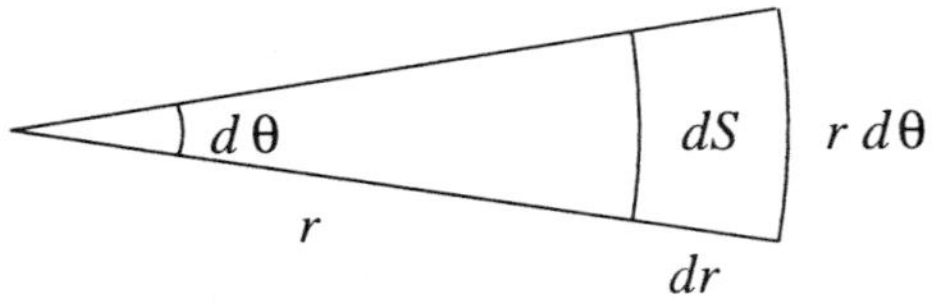

Fig. 2.8. The area element dS in polar coordinates.

element dS is almost rectangular, with length dr in the r direction and $r\,d\theta$ in the θ direction, so $dS = r\,d\theta\,dr$. Suppose now that the radius of the surface S is 1 and that $U_0 = 1 - r^2$. The value of the surface integral is then

$$Q = \iint_S U_0\,dS = \int_0^1 \int_0^{2\pi} (1 - r^2) r\,d\theta\,dr.$$

The inner, θ, integral is carried out first, with r temporarily regarded as a constant. Since there is no dependence on θ in the integral, this inner integral just gives a factor of 2π, so

$$Q = \int_0^1 2\pi(1 - r^2) r\,dr = 2\pi \left[r^2/2 - r^4/4\right]_0^1 = \pi/2.$$

Finally, consider the case where the surface S is curved. The surface can be written in terms of two parameters, v and w, so that a position vector $\boldsymbol{r}$ lying in the surface is written $\boldsymbol{r} = \boldsymbol{r}(v, w)$. Now consider a small change in the value of v, to $v + dv$. The vector $\boldsymbol{r}(v + dv, w)$ also lies on the surface, so the difference between these two vectors, $\boldsymbol{r}(v + dv, w) - \boldsymbol{r}(v, w) = (\partial\boldsymbol{r}/\partial v)\,dv$ must be a vector lying in, or tangent to, the surface, and similarly for the vector $(\partial\boldsymbol{r}/\partial w)\,dw$. To evaluate the surface integral we need an expressions for $\boldsymbol{n}\,dS$, but we know from Section 1.3 that this is simply the cross product of the two vectors, since the cross product of two vectors gives a vector perpendicular to both and with a magnitude equal to the area of the parallelogram created by the two vectors. Therefore the surface integral can be written as

$$\iint_S \boldsymbol{u}\cdot\boldsymbol{n}\,dS = \iint_S \boldsymbol{u}\cdot\frac{\partial\boldsymbol{r}}{\partial v}\times\frac{\partial\boldsymbol{r}}{\partial w}\,dv\,dw.$$

For example, consider the integral of the vector field $\boldsymbol{u} = (x, z, -y)$ over the curved surface of the cylinder $x^2 + y^2 = 1$ lying between $z = 0$ and $z = 1$. The

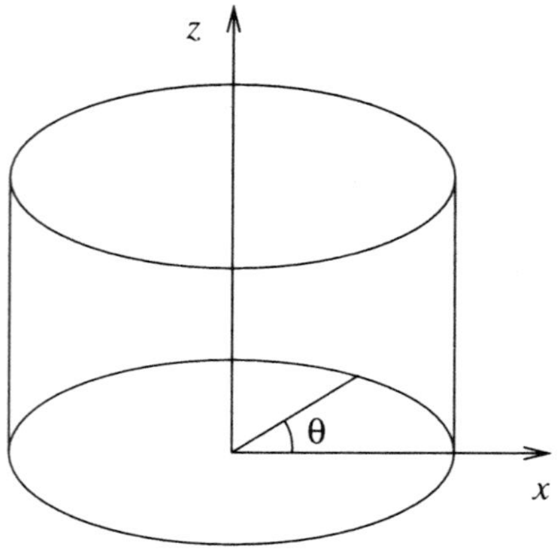

Fig. 2.9. A point on the cylindrical surface $x^2 + y^2 = 1$ can be denoted by the parameters z and θ.

two parameters describing the surface are the height z and the angle θ around the cylinder (Figure 2.9). In terms of these parameters the position vector is $\boldsymbol{r} = (x, y, z) = (\cos\theta, \sin\theta, z)$, so

$$\frac{\partial \boldsymbol{r}}{\partial \theta} = (-\sin\theta, \cos\theta, 0), \qquad \frac{\partial \boldsymbol{r}}{\partial z} = (0, 0, 1)$$

$$\text{and} \qquad \frac{\partial \boldsymbol{r}}{\partial \theta} \times \frac{\partial \boldsymbol{r}}{\partial z} = (\cos\theta, \sin\theta, 0).$$

The value of the integral is therefore

$$\begin{aligned} \iint_S \boldsymbol{u} \cdot \boldsymbol{n}\, dS &= \int_0^1 \int_0^{2\pi} (\cos\theta, z, -\sin\theta) \cdot (\cos\theta, \sin\theta, 0)\, d\theta\, dz \\ &= \int_0^1 \int_0^{2\pi} \cos^2\theta + z\sin\theta\, d\theta\, dz \\ &= \int_0^1 \pi\, dz = \pi. \end{aligned}$$

Example 2.14

Evaluate the surface integral of $\boldsymbol{u} = (y, x^2, z^2)$, over the surface S, where S is the triangular surface on $x = 0$ with $y \geq 0$, $z \geq 0$, $y + z \leq 1$, with the normal $\boldsymbol{n}$ directed in the positive x direction.

In this example $\boldsymbol{n} = (1, 0, 0)$ and so $\boldsymbol{u} \cdot \boldsymbol{n} = y$. When the surface is not rectangular, care must be taken when setting the limits of the integration. If we choose to do the z integral first, then for any given value of y, the range

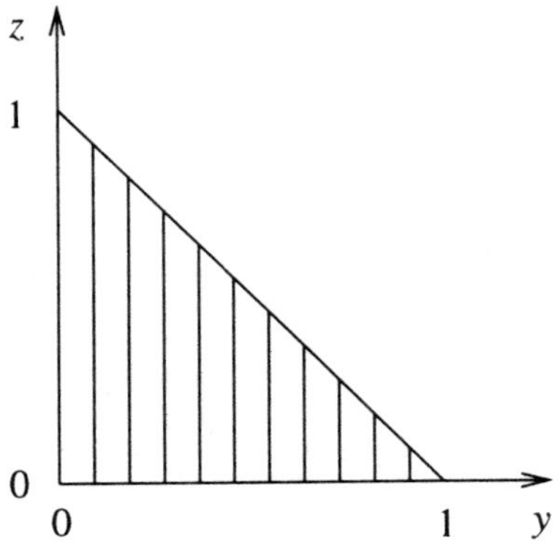

Fig. 2.10. For each value of y, z runs from the line $z = 0$ up to the line $y + z = 1$.

of values for z is $0 \le z \le 1 - y$ (Figure 2.10). The outer y integral then has limits of 0 and 1. This corresponds to covering the triangular area with vertical strips. The value of the integral is

$$\int_0^1 \int_0^{1-y} y\, dz\, dy = \int_0^1 [yz]_0^{1-y}\, dy = \int_0^1 y - y^2\, dy = 1/6.$$

The integral could also have been evaluated by doing the y integral first, in which case the limits for y are $0 \le y \le 1 - z$ and the limits for z are $0 \le z \le 1$. This ordering corresponds to covering the region of integration in Figure 2.10 with horizontal strips. The value of the integral is the same:

$$\int_0^1 \int_0^{1-z} y\, dy\, dz = \int_0^1 [y^2/2]_0^{1-z}\, dz = \int_0^1 (1-z)^2/2\, dz = 1/6.$$

Example 2.15

Find the surface integral of $\boldsymbol{u} = \boldsymbol{r}$ over the part of the paraboloid $z = 1 - x^2 - y^2$ with $z > 0$, with the normal pointing upwards.

Since the surface is curved, a description of the surface in terms of two parameters is needed. Using simply x and y, a point on the surface is $(x, y, 1 - x^2 - y^2)$ and the two tangent vectors in the surface, obtained by differentiating with respect to x and y, are $(1, 0, -2x)$ and $(0, 1, -2y)$. Taking the cross product of these two vectors, the quantity $\boldsymbol{n}\, dS$ is $(2x, 2y, 1)\, dx\, dy$. Note that this has a positive z component, so is directed upwards as required. Taking the dot product with $\boldsymbol{u}$ gives $\boldsymbol{u} \cdot \boldsymbol{n}\, dS = 2x^2 + 2y^2 + z\, dx\, dy = 1 + x^2 + y^2\, dx\, dy$. The limits on the integral are determined in a similar manner to Example 2.14. The edge of the surface is given as $z = 0$, which is the circle $x^2 + y^2 = 1$. Choosing

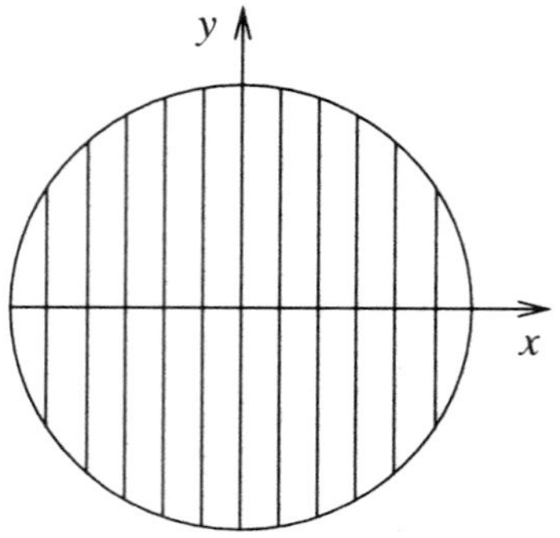

Fig. 2.11. For any value of x, y ranges from the lower half of the circle $x^2 + y^2 = 1$ to the upper half of the circle.

to do the y integral first, the y integration is carried out with x fixed, so the range of values for y is $-\sqrt{1-x^2} < y < \sqrt{1-x^2}$ (Figure 2.11). The range of values for x is $-1 < x < 1$. The value of the integral is

$$\begin{aligned}\int_{-1}^{1}\int_{-\sqrt{1-x^2}}^{\sqrt{1-x^2}} 1 + x^2 + y^2 \, dy \, dx &= \int_{-1}^{1} \left[y + x^2 y + y^3/3\right]_{-\sqrt{1-x^2}}^{\sqrt{1-x^2}} dx \\ &= \int_{-1}^{1} (8/3 + 4/3\, x^2)\sqrt{1-x^2}\, dx \\ &= 4\pi/3 + \pi/6 = 3\pi/2,\end{aligned}$$

where the final integral has been evaluated using the results of Section 2.1.2.

2.3.3 Other forms of surface integrals

The surface integral of $\boldsymbol{u} \cdot \boldsymbol{n}$ is the most important type of surface integral. However, as in the case of line integrals, other types of surface integrals can be defined, for example

$$\iint_S f \, dS, \qquad \iint_S f \, \boldsymbol{n} \, dS \qquad \text{and} \qquad \iint_S \boldsymbol{v} \times \boldsymbol{n} \, dS,$$

where f is a scalar field and $\boldsymbol{v}$ is a vector field. These integrals are evaluated using the methods of the previous section.

Example 2.16

If S is the entire x, y plane, evaluate the integral

$$I = \iint_S \mathrm{e}^{-x^2-y^2}\, dS,$$

by transforming the integral into polar coordinates.

In polar coordinates (r, θ), $x^2 + y^2 = r^2$ and $dS = r\, d\theta\, dr$. The ranges of the variables to cover the whole plane are $0 \le r < \infty$ and $0 \le \theta < 2\pi$, so

$$I = \int_0^\infty \int_0^{2\pi} \mathrm{e}^{-r^2} r\, d\theta\, dr = \int_0^\infty 2\pi \mathrm{e}^{-r^2} r\, dr = \pi \left[-\mathrm{e}^{-r^2}\right]_0^\infty = \pi.$$

This answer can be used to show that $\int_{-\infty}^{\infty} \mathrm{e}^{-x^2}\, dx = \sqrt{\pi}$, a result which cannot be obtained by standard methods of integration.

2.4 Volume integrals

2.4.1 Introductory example: mass of an object with variable density

Suppose that an object of volume V has a density ρ. If ρ is a constant, the mass M of the object is simply $M = \rho V$. Now suppose that the object has a density which is a function of position, $\rho = \rho(\boldsymbol{r})$. What is the total mass of the object?

The argument proceeds in a similar manner to the examples of line and surface integrals. The volume V is divided into N small pieces with volumes δV_i, $i = 1, \ldots, N$, which are called *volume elements*. Within each of the volume elements the density is approximately constant (assuming that ρ is a continuous function of position) and so the mass M_i of the volume element at position $\boldsymbol{r}_i$ is $M_i = \rho(\boldsymbol{r}_i)\, \delta V_i$. The total mass of the object is the sum of the masses of all the volume elements,

$$M = \sum_{i=1}^{N} \rho(\boldsymbol{r}_i)\, \delta V_i.$$

The *volume integral* of ρ over the volume V is defined to be the limit of this sum as $N \to \infty$:

$$\iiint_V \rho\, dV = \lim_{N\to\infty} \sum_{i=1}^{N} \rho(\boldsymbol{r}_i)\, \delta V_i. \tag{2.12}$$

Volume integrals can also be used to compute the volumes of objects, in which case $\rho = 1$ in the above example. Note that volume integrals usually occur as integrals of scalar quantities. However, the volume integral of a vector field $\boldsymbol{u}$,

$$\iiint_V \boldsymbol{u}\, dV$$

can be defined in a similar way.

2.4.2 Evaluation of volume integrals

Volume integrals are evaluated by carrying out three successive integrals. The same rule for the evaluation of the triple integral applies as for double integrals: the inner integral is evaluated first. The main difficulty in this process is in determining the correct limits for the integrals when the shape of the object is complicated. It is often helpful to sketch the region of integration in order to find the limits on the integrals. Also useful is the rule that in general, the limits on an integral can depend only on the variables of integrals that lie outside that integral. For example, if the integrals are evaluated in the order x, y, z then the limits on the y integral may depend on z but not on x.

Example 2.17

A cube $0 \leq x, y, z, \leq 1$ has a variable density given by $\rho = 1 + x + y + z$. What is the total mass of the cube?

The total mass is

$$\begin{aligned}
M &= \iiint_V \rho\, dV \\
&= \int_0^1 \int_0^1 \int_0^1 1 + x + y + z\, dx\, dy\, dz \\
&= \int_0^1 \int_0^1 \left[x + x^2/2 + xy + xz\right]_0^1 dy\, dz \\
&= \int_0^1 \int_0^1 (3/2 + y + z)\, dy\, dz \\
&= \int_0^1 \left[3y/2 + y^2/2 + yz\right]_0^1 dz \\
&= \int_0^1 (2 + z)\, dz \\
&= \left[2z + z^2/2\right]_0^1 = 5/2.
\end{aligned}$$

Note that in this example the integrals were carried out in the order x, y, z, but any other choice of ordering is equally valid.

Example 2.18

Find the volume of the tetrahedron with vertices at $(0,0,0)$, $(a,0,0)$, $(0,b,0)$ and $(0,0,c)$.

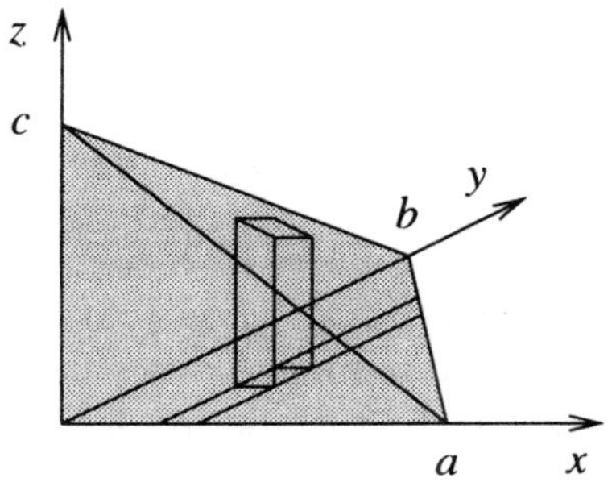

Fig. 2.12. For any given values of x and y, z ranges from the plane $z=0$ up to the plane $x/a+y/b+z/c=1$. This is indicated by the vertical column.

A sketch of the tetrahedron is shown in Figure 2.12. The faces of the tetrahedron are the planes $x=0$, $y=0$, $z=0$ and the plane which passes through the three points $(a,0,0)$, $(0,b,0)$ and $(0,0,c)$. The equation of this plane is $x/a+y/b+z/c=1$, which can be deduced from the general formula for the equation of a plane (1.3). Suppose that we choose to do the z integral first. This integral is carried out for fixed values of x and y, so the range of z is from the plane $z=0$ to the plane $z=c(1-x/a-y/b)$. Choosing to do the y integral next, y ranges from 0 to the line that passes through $(a,0,0)$ and $(0,b,0)$, which is $y=b(1-x/a)$. Finally the range of x is from 0 to a. The volume V is therefore

$$\begin{aligned} V &= \int_0^a \int_0^{b(1-x/a)} \int_0^{c(1-x/a-y/b)} dz\,dy\,dx \\ &= \int_0^a \int_0^{b(1-x/a)} c(1-x/a-y/b)\,dy\,dx \\ &= c\int_0^a \left[y(1-x/a)-y^2/2b\right]_0^{b(1-x/a)} dx \\ &= \frac{cb}{2}\int_0^a (1-x/a)^2\,dx = \frac{abc}{6}. \end{aligned}$$

Note that this result is consistent with the formula for the volume of a tetrahedron given in terms of the scalar triple product in Section 1.4.

Summary of Chapter 2

- There are several different types of integral, but each one should be interpreted as the limit of a sum.
- A *line integral*, written

$$\int_C \boldsymbol{F} \cdot d\boldsymbol{r},$$

  represents the sum of the elements $\boldsymbol{F} \cdot d\boldsymbol{r}$ along the curve C. Applications include the total amount of work done when a particle moves in the presence of a force that is a function of position.
- Line integrals are evaluated by writing the vector $\boldsymbol{F}$ and the curve C in terms of a parameter, t.
- If the line integral of $\boldsymbol{F}$ around any closed curve is zero, $\boldsymbol{F}$ is said to be *conservative*.
- The *surface integral*,

$$\iint_S \boldsymbol{u} \cdot \boldsymbol{n}\, dS,$$

  represents the flux of $\boldsymbol{u}$ across the surface S; this can be thought of as the volume of fluid flowing with velocity $\boldsymbol{u}$ across the surface S per unit time. The *normal vector* $\boldsymbol{n}$ is a unit vector that is perpendicular to the surface S.
- Surface integrals are evaluated by carrying out two successive integrations.
- The *volume integral*,

$$\iiint_V \rho\, dV,$$

  represents the sum of $\rho\, dV$ over all the volume elements dV contained within V. If ρ is the density, the volume integral gives the total mass of the object with volume V.
- Volume integrals are evaluated by carrying out three successive integrations. Care must be taken over setting the limits of the integrals and over the order in which the three integrals are evaluated.
- In both surface and volume integrals, the inner integral is evaluated first.
- There are other forms of line, surface and volume integrals, but the forms displayed above are the most commonly occurring.

EXERCISES

2.5 Evaluate the surface integral of $\boldsymbol{u} = (xy, x, x+y)$ over the surface S defined by $z = 0$ with $0 \le x \le 1$, $0 \le y \le 2$, with the normal $\boldsymbol{n}$ directed in the positive z direction.

2.6 Find the surface integral of $\boldsymbol{u} = \boldsymbol{r}$ over the surface of the unit cube $0 \le x, y, z \le 1$, with $\boldsymbol{n}$ pointing outward.

2.7 The surface S is defined to be that part of the plane $z = 0$ lying between the curves $y = x^2$ and $x = y^2$. Find the surface integral of $\boldsymbol{u} \cdot \boldsymbol{n}$ over S where $\boldsymbol{u} = (z, xy, x^2)$ and $\boldsymbol{n} = (0, 0, 1)$.

2.8 Find the surface integral of $\boldsymbol{u} \cdot \boldsymbol{n}$ over S where S is the part of the surface $z = x + y^2$ with $z < 0$ and $x > -1$, $\boldsymbol{u}$ is the vector field $\boldsymbol{u} = (2y + x, -1, 0)$ and $\boldsymbol{n}$ has a negative z component.

2.9 Find the volume integral of the scalar field $\phi = x^2 + y^2 + z^2$ over the region V specified by $0 \le x \le 1$, $1 \le y \le 2$, $0 \le z \le 3$.

2.10 Find the volume of the section of the cylinder $x^2 + y^2 = 1$ that lies between the planes $z = x + 1$ and $z = -x - 1$.

2.11 A circular pond with radius 1 m and a maximum depth of 1 m has the shape of a paraboloid, so that its depth z is $z = 1 - x^2 - y^2$. What is the total volume of the pond? How does this compare with the case where the pond has the same radius and depth but has the shape of a hemisphere?

3
Gradient, Divergence and Curl

3.1 Partial differentiation and Taylor series

This chapter introduces important concepts concerning the differentiation of scalar and vector quantities in three dimensions. These concepts form the core of the subject of vector calculus. In this preliminary section, the methods of partial differentiation and Taylor series are reviewed.

3.1.1 Partial differentiation

Consider a scalar quantity f which is a function of three variables, so $f = f(x, y, z)$. Then the *partial derivative* of f with respect to x is defined to be the derivative of f with respect to x, regarding y and z as constants. To indicate that f is a function of more than one variable, the partial derivative is written using a curly d, ∂. More formally, the definition of the partial derivative is

$$\frac{\partial f}{\partial x} = \lim_{\delta x \to 0} \frac{f(x + \delta x, y, z) - f(x, y, z)}{\delta x}. \tag{3.1}$$

Second derivatives, such as $\partial^2 f/\partial x^2$ and mixed derivatives, such as $\partial^2 f/\partial y \partial x$ can also be defined: the mixed derivative means that f is differentiated with respect to x regarding y as a constant, and then differentiated with respect to y regarding x as a constant. An important property of this mixed derivative, or cross-derivative, is that the order of the two derivatives does not matter, i.e.

$$\frac{\partial^2 f}{\partial x \partial y} = \frac{\partial^2 f}{\partial y \partial x}, \tag{3.2}$$

provided that these second partial derivatives exist and are continuous.

The main applications of partial differentiation are in the following sections. One additional application is in finding the maximum or minimum of a function of more than one variable: when a function is at a maximum or a minimum, all of its partial derivatives are zero.

Example 3.1

The function $f(x, y, z)$ is defined by $f = x^2 + xy \sin z - yz$. Find the partial derivatives of f with respect to x, y and z and verify the result (3.2) that the order of partial differentiation does not matter.

Differentiating f with respect to x, y and z in turn gives the three partial derivatives

$$\frac{\partial f}{\partial x} = 2x + y \sin z, \qquad \frac{\partial f}{\partial y} = x \sin z - z, \qquad \frac{\partial f}{\partial z} = xy \cos z - y.$$

Differentiating $\partial f / \partial y$ with respect to x gives

$$\frac{\partial^2 f}{\partial x \partial y} = \frac{\partial}{\partial x}\left(\frac{\partial f}{\partial y}\right) = \sin z,$$

and similarly, differentiating $\partial f / \partial x$ with respect to y gives

$$\frac{\partial^2 f}{\partial y \partial x} = \frac{\partial}{\partial y}\left(\frac{\partial f}{\partial x}\right) = \sin z.$$

In the same way it can be confirmed that the ordering of the cross-derivatives in x and z or in y and z does not matter.

Example 3.2

A rectangular box has height a, length b and breadth c, and is open at the top. If the volume of the box is fixed, deduce how a, b and c should be related to minimise the surface area of the box.

Let the fixed volume of the box be V, so $abc = V$. Two sides of the box have area ab, two have area ac and the base has area bc, so the total surface area is $A = 2ab + 2ac + bc$. Using the constraint $abc = V$, a can be eliminated so that $A = 2V/c + 2V/b + bc$. At a maximum or a minimum, the partial derivatives of A with respect to b and c must both be zero. This gives the two simultaneous equations

$$-2V/b^2 + c = 0, \qquad -2V/c^2 + b = 0.$$

From the first equation, $c = 2V/b^2$ and substituting this into the second equation gives the solution for b, $b^3 = 2V$ or $b = (2V)^{1/3}$, and similarly $c = (2V)^{1/3}$. The condition $abc = V$ gives $a = V/((2V)^{2/3}) = (2V)^{1/3}/2$. Therefore in the arrangement that minimises the surface area, $b = c$ and $a = b/2$, so the height of the box is half its length.

Note that we have not shown that this is a minimum and not a maximum. However, common sense suggests that since the area would be very large if the box were tall and thin, the solution found probably does represent a minimum.

3.1.2 Taylor series in more than one variable

The Taylor series for an infinitely differentiable function $f(x)$ of a single variable is

$$\begin{aligned} f(x) &= f(a) + (x-a)\frac{df}{dx}(a) + \frac{(x-a)^2}{2!}\frac{d^2f}{dx^2}(a) + \dots \\ &= \sum_{n=0}^{\infty} \frac{(x-a)^n}{n!}\frac{d^n f}{dx^n}(a). \end{aligned} \tag{3.3}$$

This can also be written as

$$\delta f = \delta x \frac{df}{dx} + \frac{(\delta x)^2}{2!}\frac{d^2 f}{dx^2} + \dots, \tag{3.4}$$

where $\delta x = (x - a)$ is a small perturbation and $\delta f = f(x) - f(a)$ is the corresponding perturbation in the value of the function.

Taylor series can also be constructed for functions of more than one variable. For a function $f(x, y)$ of two independent variables, the analogous formula is

$$\delta f = \delta x \frac{\partial f}{\partial x} + \delta y \frac{\partial f}{\partial y} + \frac{(\delta x)^2}{2!}\frac{\partial^2 f}{\partial x^2} + \frac{(\delta y)^2}{2!}\frac{\partial^2 f}{\partial y^2} + \delta x \delta y \frac{\partial^2 f}{\partial x \partial y} + \dots. \tag{3.5}$$

In the following sections we will make use of the Taylor series for a function $f(x, y, z)$ of three variables, but in all cases only the linear terms, that is, only those that only involve a single power of δx, δy or δz, will be needed:

$$\delta f = \delta x \frac{\partial f}{\partial x} + \delta y \frac{\partial f}{\partial y} + \delta z \frac{\partial f}{\partial z} + \dots. \tag{3.6}$$

Taylor series can be useful for approximating functions, as in the following example.

Example 3.3

Find an approximate value for the function $f(x, y, z) = 2x + (1 + y)\sin z$ at the point $x = 0.1$, $y = 0.2$, $z = 0.3$.

The function f takes the value 0 at the point $(0, 0, 0)$. Near to this point, the function can be approximated by its Taylor series expansion. To do this, the three partial derivatives of f evaluated at $(0, 0, 0)$ are required. These are

$$\frac{\partial f}{\partial x} = 2, \qquad \frac{\partial f}{\partial y} = \sin z = 0, \qquad \frac{\partial f}{\partial z} = (1 + y)\cos z = 1.$$

Hence the Taylor expansion (3.6) is

$$\delta f = 2\delta x + \delta z + \ldots,$$

which at the point $(0.1, 0.2, 0.3)$ gives the approximate value $f \approx 0.5$ (the correct value to four decimal places is 0.5546).

3.2 Gradient of a scalar field

In Section 1.6 the concept of a scalar field was introduced as a scalar quantity which is a function of position in space. A scalar field f can be visualised in terms of the ***level surfaces*** or ***isosurfaces*** on which f is constant. The ***gradient*** of the scalar field f is a vector field, with a direction that is perpendicular to the level surfaces, pointing in the direction of increasing f, with a magnitude equal to the rate of change of f in this direction (Figure 3.1).

The gradient of a scalar field f can be written as $\operatorname{grad} f$, but the gradient is so important that a special symbol for grad, $\boldsymbol{\nabla}$, is used, so $\operatorname{grad} f = \boldsymbol{\nabla} f$. This symbol is sometimes referred to as 'del' or 'nabla'.

The gradient of f can also be defined in a Cartesian coordinate system in terms of the partial derivatives of f:

$$\boldsymbol{\nabla} f = \frac{\partial f}{\partial x}\mathbf{e}_1 + \frac{\partial f}{\partial y}\mathbf{e}_2 + \frac{\partial f}{\partial z}\mathbf{e}_3. \tag{3.7}$$

We will now show that these two definitions are equivalent, by showing that the vector $\boldsymbol{\nabla} f$ defined in (3.7) satisfies the two conditions of being perpendicular to the level surfaces and with magnitude equal to the rate of change of f in this direction.

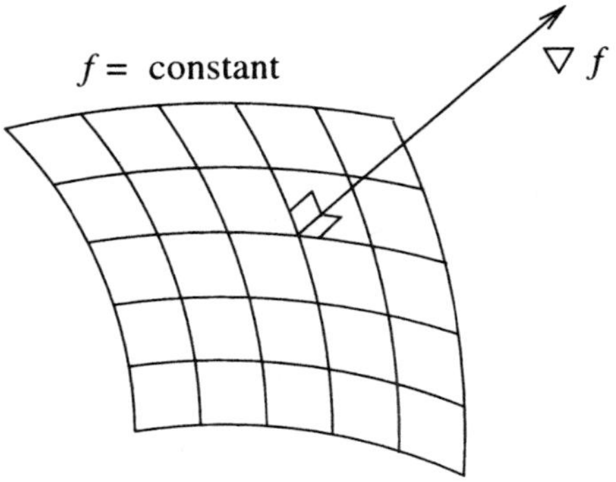

Fig. 3.1. The gradient of f is a vector perpendicular to the surface $f =$ constant.

Consider an infinitesimal change in position in space from $\boldsymbol{r}$ to $\boldsymbol{r} + \boldsymbol{dr}$. This results in a small change in the value of the scalar field f from f to $f + df$, where, from (3.6),

$$\begin{aligned} df &= \frac{\partial f}{\partial x}dx + \frac{\partial f}{\partial y}dy + \frac{\partial f}{\partial z}dz \\ &= \left(\frac{\partial f}{\partial x}, \frac{\partial f}{\partial y}, \frac{\partial f}{\partial z}\right) \cdot (dx, dy, dz) \\ &= \boldsymbol{\nabla} f \cdot \boldsymbol{dr}. \end{aligned} \tag{3.8}$$

Now suppose that $\boldsymbol{dr}$ lies in the surface $f =$ constant (Figure 3.2). In this case the change in the value of f must be zero, so we have $df = \boldsymbol{\nabla} f \cdot \boldsymbol{dr} = 0$. Now in general $\boldsymbol{\nabla} f \neq \boldsymbol{0}$ and $\boldsymbol{dr} \neq \boldsymbol{0}$, so the two vectors $\boldsymbol{\nabla} f$ and $\boldsymbol{dr}$ must be perpendicular. Since $\boldsymbol{dr}$ is in the level surface $f =$ constant, the vector $\boldsymbol{\nabla} f$ must be perpendicular to the level surface.

So the vector $\boldsymbol{\nabla} f$ defined by (3.7) has the correct direction; it remains to be shown that it has the correct magnitude. This is achieved by using (3.8) with $\boldsymbol{dr} = \boldsymbol{n}\, ds$, where $\boldsymbol{n}$ is the unit normal to the level surface and s is a distance measured along the normal. In this case, $df = \boldsymbol{\nabla} f \cdot \boldsymbol{n}\, ds = |\boldsymbol{\nabla} f|\, ds$, since $\boldsymbol{\nabla} f$ and $\boldsymbol{n}$ are parallel and $|\boldsymbol{n}| = 1$. Hence the magnitude of $\boldsymbol{\nabla} f$ is

$$|\boldsymbol{\nabla} f| = \frac{df}{ds}, \tag{3.9}$$

which is the rate of change of f with position along the normal.

From (3.8) it follows that $\boldsymbol{\nabla} f$ can be used to find the rate of change of f in any direction. To find the rate of change of f in the direction of the unit vector $\boldsymbol{u}$, set $\boldsymbol{dr} = \boldsymbol{u}\, ds$ where ds is the distance along $\boldsymbol{u}$. Then $df = \boldsymbol{\nabla} f \cdot \boldsymbol{u}\, ds$ and so

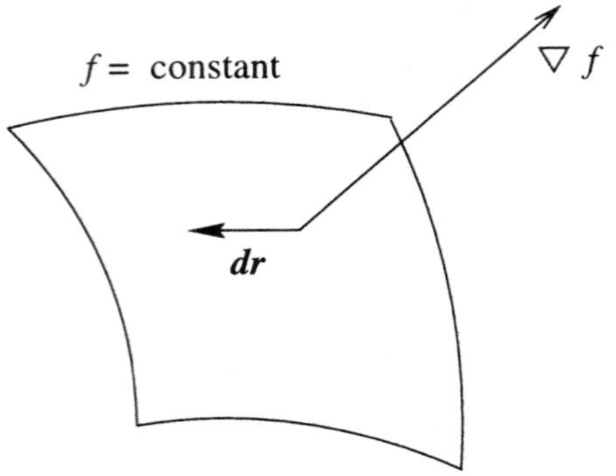

Fig. 3.2. Grad f is perpendicular to any vector $\boldsymbol{dr}$ lying in the surface f = constant.

$$\frac{df}{ds} = \boldsymbol{\nabla} f \cdot \boldsymbol{u}. \tag{3.10}$$

This is the rate of change of f in the direction of the unit vector $\boldsymbol{u}$, and is called the *directional derivative* of f. This can also be written as

$$\frac{df}{ds} = |\boldsymbol{\nabla} f| \cos\theta, \tag{3.11}$$

where θ is the angle between $\boldsymbol{\nabla} f$ and the unit vector $\boldsymbol{u}$. Since $-1 \leq \cos\theta \leq 1$, it follows that the magnitude of $\boldsymbol{\nabla} f$ is equal to the maximum rate of change of f with position.

The symbol $\boldsymbol{\nabla}$ can be interpreted as a vector differential operator,

$$\boldsymbol{\nabla} = \left(\frac{\partial}{\partial x}, \frac{\partial}{\partial y}, \frac{\partial}{\partial z}\right), \tag{3.12}$$

where the term *operator* means that $\boldsymbol{\nabla}$ only has a meaning when it acts on some other quantity.

The gradient has many important applications. These include finding normals to surfaces and obtaining the rates of change of functions in any direction, as in the following examples.

Example 3.4

Find the unit normal $\boldsymbol{n}$ to the surface $x^2 + y^2 - z = 0$ at the point $(1, 1, 2)$.

Define $f(x, y, z) = x^2 + y^2 - z = 0$, so the surface is $f = 0$. Then $\boldsymbol{\nabla} f = (2x, 2y, -1)$. At the point $(1, 1, 2)$, $\boldsymbol{\nabla} f = (2, 2, -1)$. This is a vector normal to the surface. To find the unit normal we need to divide by the magnitude, which is $(2^2 + 2^2 + 1^2)^{1/2} = 3$ so $\boldsymbol{n} = \boldsymbol{\nabla} f / |\boldsymbol{\nabla} f| = (2/3, 2/3, -1/3)$. Note that the unit normal is not uniquely defined: the vector $-\boldsymbol{n} = (-2/3, -2/3, 1/3)$ is also a unit normal to the surface.

Example 3.5

Find the directional derivative of the scalar field $f = 2x + y + z^2$ in the direction of the vector (1,1,1), and evaluate this at the origin.

The gradient of f is $\boldsymbol{\nabla} f = (2, 1, 2z)$. To find the directional derivative, we must take the dot product with the unit vector in the direction of $(1, 1, 1)$ which is $\boldsymbol{u} = (1, 1, 1)/\sqrt{3}$. The directional derivative is then $\boldsymbol{\nabla} f \cdot \boldsymbol{u} = (3 + 2z)/\sqrt{3}$. At the origin, $x = y = z = 0$, the directional derivative takes the value $\sqrt{3}$.

3.2.1 Gradients, conservative fields and potentials

There is a very important link between the gradient of a scalar field and the concept of a conservative vector field defined in Section 2.2.3. Recall that a conservative vector field is one in which the line integral along a curve connecting two points does not depend on the path taken. The connection between gradients and conservative fields is given by the following theorem.

Theorem 3.1

Suppose that a vector field $\boldsymbol{F}$ is related to a scalar field ϕ by $\boldsymbol{F} = \boldsymbol{\nabla}\phi$ and $\boldsymbol{\nabla}\phi$ exists everywhere in some region D. Then $\boldsymbol{F}$ is conservative within D. Conversely, if $\boldsymbol{F}$ is conservative, then $\boldsymbol{F}$ can be written as the gradient of a scalar field, $\boldsymbol{F} = \boldsymbol{\nabla}\phi$.

Proof

Suppose that $\boldsymbol{F} = \boldsymbol{\nabla}\phi$. Then the line integral of $\boldsymbol{F}$ along a curve C connecting two points A and B is

$$\int_C \boldsymbol{F} \cdot d\boldsymbol{r} = \int_C \boldsymbol{\nabla}\phi \cdot d\boldsymbol{r}.$$

Using (3.8) this can be written as

$$\int_C \boldsymbol{F} \cdot d\boldsymbol{r} = \int_C d\phi = [\phi]_A^B = \phi(B) - \phi(A),$$

where the line integral has been evaluated simply using ϕ as the parameter. Since this result only depends on the end points of C, $\boldsymbol{F}$ is conservative.

Conversely, suppose that $\boldsymbol{F}$ is conservative. Then a scalar field $\phi(\boldsymbol{r})$ can be defined as the line integral of $\boldsymbol{F}$ from the origin to the point $\boldsymbol{r}$:

$$\phi(\boldsymbol{r}) = \int_{\boldsymbol{0}}^{\boldsymbol{r}} \boldsymbol{F} \cdot d\boldsymbol{r}. \tag{3.13}$$

Since $\boldsymbol{F}$ is conservative, the value of ϕ does not depend on the path taken from $\boldsymbol{0}$ to $\boldsymbol{r}$, so ϕ is well defined. From the definition of an integral it then follows that an infinitesimal change in ϕ is given by

$$d\phi = \boldsymbol{F} \cdot \boldsymbol{dr}.$$

Comparing this with (3.8) shows that $\boldsymbol{F} \cdot \boldsymbol{dr} = \boldsymbol{\nabla}\phi \cdot \boldsymbol{dr}$. This must be true for any choice of $\boldsymbol{dr}$ and so $\boldsymbol{F} = \boldsymbol{\nabla}\phi$. □

If a vector field $\boldsymbol{F}$ is conservative, the corresponding scalar field ϕ which obeys $\boldsymbol{F} = \boldsymbol{\nabla}\phi$ is called the *potential* for $\boldsymbol{F}$. Note that the potential is not unique, since an arbitrary constant can be added to ϕ without affecting $\boldsymbol{\nabla}\phi$. This arbitrary constant corresponds to the arbitrary choice of the origin for the lower limit in the integral in the definition (3.13).

Example 3.6

Show that the vector field $\boldsymbol{F} = (2x + y, x, 2z)$ is conservative.

$\boldsymbol{F}$ is conservative if it can be written as the gradient of a scalar field ϕ. This gives the three equations

$$\frac{\partial \phi}{\partial x} = 2x + y, \qquad \frac{\partial \phi}{\partial y} = x, \qquad \frac{\partial \phi}{\partial z} = 2z.$$

Integrating the first of these equations with respect to x gives $\phi = x^2 + xy + h(y, z)$ where h is an arbitrary function of y and z, analogous to a constant of integration. The second equation forces the partial derivative of h with respect to y to be zero, so that h only depends on z. The third equation yields $dh/dz = 2z$, so $h(z) = z^2 + c$, where c is any constant. Therefore all three equations are satisfied by the potential function

$$\phi = x^2 + xy + z^2$$

and $\boldsymbol{F}$ is a conservative vector field.

3.2.2 Physical applications of the gradient

The gradient of a scalar field appears in many physical contexts. Two examples are given below.

- Let p denote the pressure within a gas. Then there is a force $\boldsymbol{F}$ acting on any volume element δV due to the pressure gradient, given by $\boldsymbol{F} = -\boldsymbol{\nabla}p\,\delta V$.
- A material has a constant thermal conductivity K and a variable temperature $T(\boldsymbol{r})$. Because of the temperature variation, heat flows from the hot regions to the cold regions. The heat flux $\boldsymbol{q}$ is a vector quantity, $\boldsymbol{q} = -K\boldsymbol{\nabla}T$.

EXERCISES

3.1 Find the gradient of the scalar field $f = xyz$, and evaluate it at the point (1,2,3). Hence find the directional derivative of f at this point in the direction of the vector $(1,1,0)$.

3.2 Find the unit normal to the surface $y = x + z^3$ at the point $(1,2,1)$.

3.3 Show that the gradient of the scalar field $\phi = r = |\boldsymbol{r}|$ is $\boldsymbol{r}/r$ and interpret this result geometrically.

3.4 Find the angle between the surfaces of the sphere $x^2 + y^2 + z^2 = 2$ and the cylinder $x^2 + y^2 = 1$ at a point where they intersect.

3.5 Find the gradient of the scalar field $f = yx^2 + y^3 - y$ and hence find the minima and maxima of f. Sketch the contours $f =$ constant and the vector field $\boldsymbol{\nabla} f$.

3.6 If $\boldsymbol{a}$ is a constant vector, find the gradient of $f = \boldsymbol{a} \cdot \boldsymbol{r}$ and interpret this result geometrically.

3.7 Determine whether or not the vector field $\boldsymbol{F} = (\sin y, x, 0)$ is conservative.

3.8 Consider the vector field $\boldsymbol{F} = (y/(x^2 + y^2), -x/(x^2 + y^2), 0)$. Show that $\boldsymbol{F}$ can be written as the gradient of a potential ϕ. Show also that the line integral of $\boldsymbol{F}$ around the unit circle $x^2 + y^2 = 1$ is non-zero. Explain why this result does not contradict Theorem 3.1.

3.3 Divergence of a vector field

This section introduces the first of two ways of differentiating a vector field, the divergence. The second way of differentiating a vector field, the curl, is defined in Section 3.4. Each of these quantities is defined in terms of an integral.

The *divergence* of a vector field $\boldsymbol{u}$ is a scalar field. Its value at a point P is defined by

$$\operatorname{div} \boldsymbol{u} = \lim_{\delta V \to 0} \frac{1}{\delta V} \oiint_{\delta S} \boldsymbol{u} \cdot \boldsymbol{n} \, dS, \tag{3.14}$$

where δV is a small volume enclosing P with surface δS and $\boldsymbol{n}$ is the outward pointing normal to δS. Physically, this corresponds to the amount of flux of the vector field $\boldsymbol{u}$ out of δV divided by the volume δV (Figure 3.3).

As in the case of the gradient, this physical definition leads to an equivalent definition in terms of the components of $\boldsymbol{u} = (u_1, u_2, u_3)$ in a Cartesian

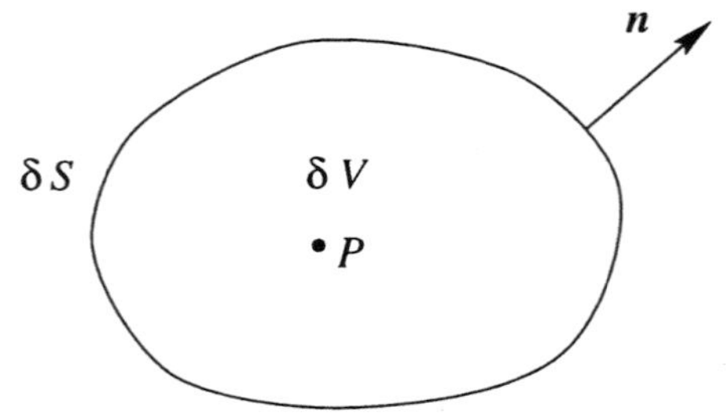

Fig. 3.3. Definition of the divergence. The small volume δV has surface δS and outward normal $\boldsymbol{n}$.

coordinate system. To derive this alternative form, take the volume δV to be a small rectangular box with sides of length δx, δy and δz and centred on a point (x, y, z) (Figure 3.4). It is assumed that the components of $\boldsymbol{u}$ have continuous partial derivatives.

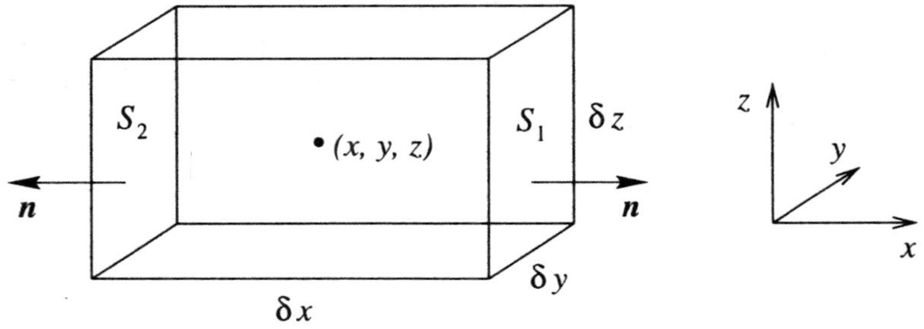

Fig. 3.4. Rectangular box used to obtain an expression for div $\boldsymbol{u}$ in Cartesian coordinates.

Since the rectangular box has six faces, there are six different contributions to the surface integral in (3.14). Consider first the contribution from the face labelled S_1 in Figure 3.4. This face is perpendicular to the x-axis, so the unit outward normal is $(1, 0, 0)$ and hence $\boldsymbol{u} \cdot \boldsymbol{n} = u_1$. The centre of face S_1 is at the point $(x + \delta x/2, y, z)$ and the area of the face is $\delta y \, \delta z$, so the contribution to the surface integral from this face is

$$\iint_{S_1} \boldsymbol{u} \cdot \boldsymbol{n} \, dS \approx u_1(x + \delta x/2, y, z) \, \delta y \, \delta z,$$

where we have used the fact that since the surface is small, the surface integral can be approximated as the value of $\boldsymbol{u} \cdot \boldsymbol{n}$ multiplied by the area of the surface.

A similar argument can be used to approximate the contribution to the surface integral from S_2, which is located at $(x - \delta x/2, y, z)$. The unit outward normal for S_2 is $(-1, 0, 0)$ so $\boldsymbol{u} \cdot \boldsymbol{n} = -u_1$ and the contribution to the surface integral is

$$\iint_{S_2} \boldsymbol{u} \cdot \boldsymbol{n}\, dS \approx -u_1(x - \delta x/2, y, z)\, \delta y\, \delta z.$$

Adding the contributions from these two surfaces and making use of the definition of the partial derivative (3.1), the combined contribution to the surface integral is

$$\begin{aligned}
\iint_{S_1+S_2} \boldsymbol{u} \cdot \boldsymbol{n}\, dS &\approx \left(u_1(x + \frac{\delta x}{2}, y, z) - u_1(x - \frac{\delta x}{2}, y, z) \right) \delta y\, \delta z \\
&\approx \frac{\partial u_1}{\partial x}\, \delta x\, \delta y\, \delta z \\
&\approx \frac{\partial u_1}{\partial x}\, \delta V.
\end{aligned}$$

Hence the contribution to $\operatorname{div} \boldsymbol{u}$ defined in (3.14) from surfaces S_1 and S_2 is $\partial u_1/\partial x$. Note that this is now exact since the divergence is defined by taking the limit $\delta V \to 0$. Similarly, the contribution to $\operatorname{div} \boldsymbol{u}$ from the two surfaces perpendicular to the y-axis is $\partial u_2/\partial y$ and that from the surfaces perpendicular to the z-axis is $\partial u_3/\partial z$. These are found simply by permuting the x-, y- and z-axes. Finally, adding all six contributions together gives the definition of $\operatorname{div} \boldsymbol{u}$ in terms of the Cartesian components of $\boldsymbol{u}$:

$$\operatorname{div} \boldsymbol{u} = \frac{\partial u_1}{\partial x} + \frac{\partial u_2}{\partial y} + \frac{\partial u_3}{\partial z}. \tag{3.15}$$

The divergence of $\boldsymbol{u}$ can also be written in terms of the differential operator $\boldsymbol{\nabla}$ defined in (3.12), since

$$\operatorname{div} \boldsymbol{u} = \frac{\partial u_1}{\partial x} + \frac{\partial u_2}{\partial y} + \frac{\partial u_3}{\partial z} = \left(\frac{\partial}{\partial x}, \frac{\partial}{\partial y}, \frac{\partial}{\partial z} \right) \cdot (u_1, u_2, u_3) = \boldsymbol{\nabla} \cdot \boldsymbol{u}. \tag{3.16}$$

The form $\boldsymbol{\nabla} \cdot \boldsymbol{u}$ will be used to indicate the divergence of $\boldsymbol{u}$ in the remainder of this book.

Example 3.7

Find the divergence of the vector field $\boldsymbol{u} = \boldsymbol{r}$.

The components of $\boldsymbol{u} = \boldsymbol{r}$ are $\boldsymbol{u} = (x, y, z)$. The divergence of $\boldsymbol{u}$ is therefore

$$\operatorname{div} \boldsymbol{u} = \frac{\partial x}{\partial x} + \frac{\partial y}{\partial y} + \frac{\partial z}{\partial z} = 3. \tag{3.17}$$

3.3.1 Physical interpretation of divergence

The physical definition of the divergence (3.14) gives an intuitive meaning in terms of the flux of the vector field out of a small closed surface. This can also be interpreted as the rate of 'expansion' or 'stretching' of the vector field. Consider for example the simple vector field $\boldsymbol{u} = (x, 0, 0)$. This vector field only has a component in the x direction and it is sketched in Figure 3.5(a). It is useful to think of vector fields as representing the motion of a gas. Figure 3.5(a) then represents a gas which is expanding, and the divergence of u, from (3.15), is 1 everywhere. The vector field $\boldsymbol{v} = -\boldsymbol{u} = (-x, 0, 0)$ is shown in Figure 3.5(b). This vector field is contracting, and its divergence is $\boldsymbol{\nabla} \cdot \boldsymbol{v} = -1$. Finally consider the vector field $\boldsymbol{w} = (0, x, 0)$, sketched in Figure 3.5(c). This vector field is neither expanding nor contracting, and its divergence is zero. A vector field $\boldsymbol{w}$ for which $\boldsymbol{\nabla} \cdot \boldsymbol{w} = 0$ everywhere is said to be *solenoidal.*

3.3.2 Laplacian of a scalar field

Suppose that a scalar field ϕ is twice differentiable. Then the gradient of ϕ is a differentiable vector field $\boldsymbol{\nabla}\phi$, so we can take the divergence of $\boldsymbol{\nabla}\phi$ and obtain another scalar field. This scalar field, $\boldsymbol{\nabla} \cdot \boldsymbol{\nabla}\phi$ is called the ***Laplacian*** of ϕ and has its own symbol, $\nabla^2\phi$, so

$$\boldsymbol{\nabla} \cdot \boldsymbol{\nabla}\phi = \nabla^2\phi.$$

The Laplacian of ϕ is often referred to as 'del squared ϕ'. The formula for $\nabla^2\phi$ can be found by combining the formulae for div (3.15) and grad (3.7),

$$\begin{aligned} \nabla^2\phi &= \frac{\partial}{\partial x}\left(\frac{\partial\phi}{\partial x}\right) + \frac{\partial}{\partial y}\left(\frac{\partial\phi}{\partial y}\right) + \frac{\partial}{\partial z}\left(\frac{\partial\phi}{\partial z}\right) \\ &= \frac{\partial^2\phi}{\partial x^2} + \frac{\partial^2\phi}{\partial y^2} + \frac{\partial^2\phi}{\partial z^2}. \end{aligned} \tag{3.18}$$

Thus the Laplacian of ϕ is just the sum of the second partial derivatives of ϕ.

The Laplacian can also act on a vector quantity, in which case the result is a vector whose components are the Laplacians of the components of the original vector:

$$\nabla^2\boldsymbol{u} = \left(\nabla^2 u_1, \nabla^2 u_2, \nabla^2 u_3\right). \tag{3.19}$$

The Laplacian is a very important quantity, occurring in many physical applications including heat transfer and wave motion. These applications will be considered in Chapter 8. The equation $\nabla^2\phi = 0$ is known as ***Laplace's equation.***

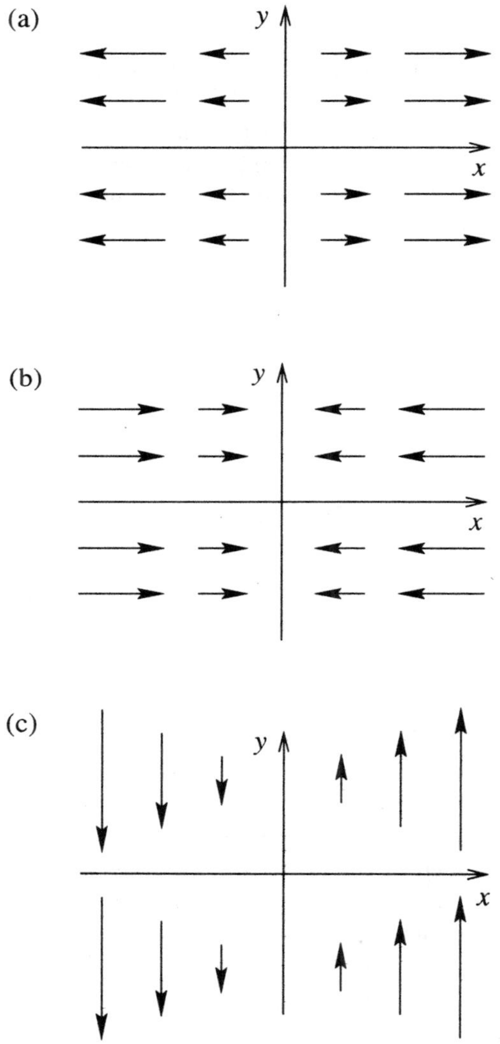

Fig. 3.5. The three vector fields (a) $\boldsymbol{u} = (x, 0, 0)$, (b) $\boldsymbol{v} = (-x, 0, 0)$, (c) $\boldsymbol{w} = (0, x, 0)$.

3.4 Curl of a vector field

The *curl* of a vector field $\boldsymbol{u}$ is a vector field. Its component in the direction of the unit vector $\boldsymbol{n}$ is

$$\boldsymbol{n} \cdot \operatorname{curl} \boldsymbol{u} = \lim_{\delta S \to 0} \frac{1}{\delta S} \oint_{\delta C} \boldsymbol{u} \cdot d\boldsymbol{r}, \tag{3.20}$$

where δS is a small surface element perpendicular to $\boldsymbol{n}$, δC is the closed curve forming the boundary of δS and δC and $\boldsymbol{n}$ are oriented in a right-handed sense, as shown in Figure 3.6. Note that this has a similar form to the definition of divergence (3.14), but with a line integral instead of a surface integral.

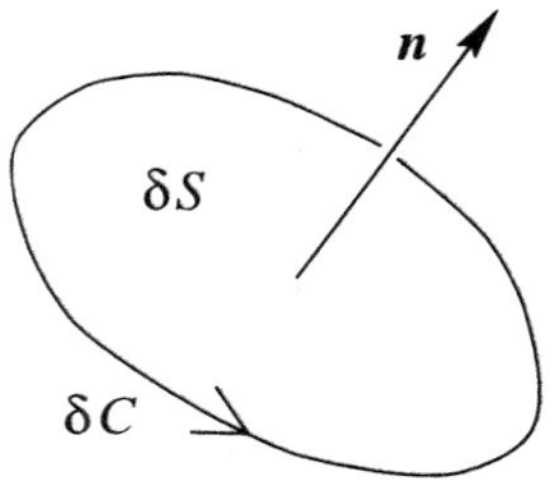

Fig. 3.6. Definition of the curl of a vector field. The small surface δS is enclosed by the curve δC and has unit normal vector $\boldsymbol{n}$.

To obtain an expression for $\operatorname{curl} \boldsymbol{u}$ in terms of the components (u_1, u_2, u_3) of $\boldsymbol{u}$, choose $\boldsymbol{n} = \boldsymbol{e_3}$, the unit vector in the z direction, to determine the z component of $\operatorname{curl} \boldsymbol{u}$. The surface δS then lies in the x, y plane and can be chosen to be a small rectangle with sides of length δx, δy centred on the point (x, y, z), with area $\delta S = \delta x\, \delta y$. The right-hand rule means that the line integral in (3.20) must be taken in the anticlockwise direction. The line integral then has four sections, as shown in Figure 3.7. It is assumed that the vector field $\boldsymbol{u}$ is differentiable with continuous partial derivatives.

Consider first the section C_1 of the line integral, which has its centre at the point $(x, y - \delta y/2, z)$. Here, the line integral is directed in the positive x direction, so $\boldsymbol{u} \cdot d\boldsymbol{r} = u_1\, dx$. Since the length δx is small, the contribution to the line integral is approximately

$$\int_{C_1} \boldsymbol{u} \cdot d\boldsymbol{r} \approx u_1(x, y - \delta y/2, z)\, \delta x.$$

Similarly, on C_3, centred at $(x, y + \delta y/2, z)$, the integral is directed in the negative x direction, so

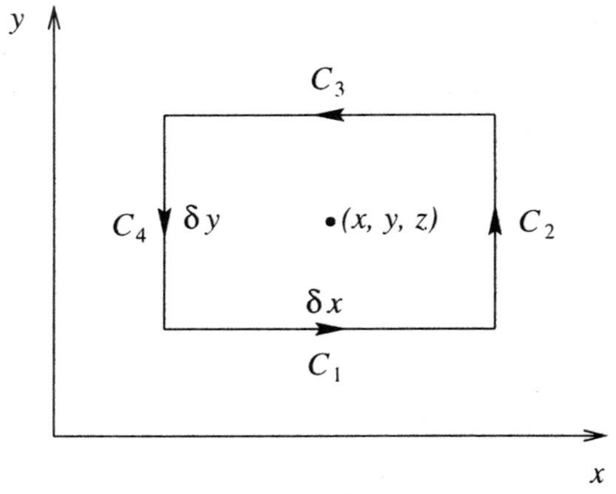

Fig. 3.7. A rectangle of four line segments is used to find an expression for curl $\boldsymbol{u}$ in Cartesian coordinates.

$$\int_{C_3} \boldsymbol{u} \cdot \boldsymbol{dr} \approx -u_1(x, y + \delta y/2, z)\, \delta x.$$

Adding these two contributions together gives

$$\begin{aligned}\int_{C_1+C_3} \boldsymbol{u} \cdot \boldsymbol{dr} &\approx (u_1(x, y - \delta y/2, z) - u_1(x, y + \delta y/2, z))\, \delta x \\ &\approx -\frac{\partial u_1}{\partial y}\, \delta y\, \delta x.\end{aligned}$$

Proceeding in a similar way with the line integrals along C_2, located at $(x + \delta x/2, y, z)$ with $\boldsymbol{dr}$ directed in the positive y direction, and C_4, located at $(x - \delta x/2, y, z)$ with $\boldsymbol{dr}$ directed in the negative y direction, we obtain

$$\begin{aligned}\int_{C_2+C_4} \boldsymbol{u} \cdot \boldsymbol{dr} &\approx (u_2(x + \delta x/2, y, z) - u_2(x - \delta x/2, y, z))\, \delta y \\ &\approx \frac{\partial u_2}{\partial x}\, \delta x\, \delta y.\end{aligned}$$

Adding together all four contributions, dividing by δS and taking the limit $\delta S \to 0$ gives the z component of curl $\boldsymbol{u}$:

$$\boldsymbol{e}_3 \cdot \operatorname{curl} \boldsymbol{u} = \frac{\partial u_2}{\partial x} - \frac{\partial u_1}{\partial y}. \tag{3.21}$$

The other components can be found by permuting x, y and z cyclically ($x \to y \to z \to x$, $u_1 \to u_2 \to u_3 \to u_1$), giving

$$\operatorname{curl}\boldsymbol{u} = \left(\frac{\partial u_3}{\partial y} - \frac{\partial u_2}{\partial z}, \frac{\partial u_1}{\partial z} - \frac{\partial u_3}{\partial x}, \frac{\partial u_2}{\partial x} - \frac{\partial u_1}{\partial y}\right). \tag{3.22}$$

Notice that there is a similarity between this formula and that for the cross product of two vectors (1.4). Curl $\boldsymbol{u}$ can also be written in terms of a determinant,

$$\operatorname{curl}\boldsymbol{u} = \begin{vmatrix} \boldsymbol{e}_1 & \boldsymbol{e}_2 & \boldsymbol{e}_3 \\ \frac{\partial}{\partial x} & \frac{\partial}{\partial y} & \frac{\partial}{\partial z} \\ u_1 & u_2 & u_3 \end{vmatrix}$$

provided that the determinant is expanded so that the partial derivatives act on the components of $\boldsymbol{u}$. This can also be written as the cross product of the differential operator $\boldsymbol{\nabla}$ and the vector $\boldsymbol{u}$, so

$$\operatorname{curl}\boldsymbol{u} = \boldsymbol{\nabla} \times \boldsymbol{u}.$$

The notation $\boldsymbol{\nabla} \times \boldsymbol{u}$ will be used henceforth.

Example 3.8

The vector field $\boldsymbol{u}$ is defined by $\boldsymbol{u} = (xy, z + x, y)$. Calculate $\boldsymbol{\nabla} \times \boldsymbol{u}$ and find the points where $\boldsymbol{\nabla} \times \boldsymbol{u} = \boldsymbol{0}$.

The components of $\boldsymbol{\nabla} \times \boldsymbol{u}$ are found using (3.22):

$$\begin{aligned}\boldsymbol{\nabla} \times \boldsymbol{u} &= \left(\frac{\partial y}{\partial y} - \frac{\partial(z+x)}{\partial z}, \frac{\partial(xy)}{\partial z} - \frac{\partial y}{\partial x}, \frac{\partial(z+x)}{\partial x} - \frac{\partial(xy)}{\partial y}\right) \\ &= (1-1, 0-0, 1-x) = (0, 0, 1-x).\end{aligned}$$

Hence $\boldsymbol{\nabla} \times \boldsymbol{u} = \boldsymbol{0}$ on the plane $x = 1$.

3.4.1 Physical interpretation of curl

From the physical definition of $\boldsymbol{\nabla} \times \boldsymbol{u}$ given in (3.20) and Figure 3.6 it is clear that $\boldsymbol{\nabla} \times \boldsymbol{u}$ is related to the rotation or twisting of the vector field $\boldsymbol{u}$.

Consider the three simple vector fields shown in Figure 3.5. For the first of these, $\boldsymbol{u} = (x, 0, 0)$, the vector field is expanding but there is no sense of rotation, and computing the curl gives $\boldsymbol{\nabla} \times \boldsymbol{u} = \boldsymbol{0}$. A vector field $\boldsymbol{u}$ for which $\boldsymbol{\nabla} \times \boldsymbol{u} = \boldsymbol{0}$ everywhere is said to be *irrotational*. Similarly, the second example, $\boldsymbol{v} = (-x, 0, 0)$, is also irrotational. For the third example, $\boldsymbol{w} = (0, x, 0)$, Figure 3.5(c), $\boldsymbol{\nabla} \times \boldsymbol{w} = (0, 0, 1)$, so there is a component of $\boldsymbol{\nabla} \times \boldsymbol{w}$ in the z direction, out of the page. The vector field $\boldsymbol{w}$ has a rotation associated with it in the following sense. Think of $\boldsymbol{w}$ as the velocity of a fluid. Then a small particle placed in this fluid will rotate in an anticlockwise sense as it moves with the fluid, since at any point the velocity component in the y direction to the right of the particle is greater than that on the left. This rotation is about an axis in the z direction, which is in the direction of $\nabla \times \boldsymbol{w}$. The vector $\nabla \times \boldsymbol{w}$ can therefore be related to the rotation of a small particle placed in the velocity field $\boldsymbol{w}$: the rate of rotation depends on the magnitude of $\nabla \times \boldsymbol{w}$, and the axis of rotation is in the direction of $\nabla \times \boldsymbol{w}$. In the context of the motion of a fluid, the curl of the velocity field is often referred to as the *vorticity* of the fluid. This relationship between rotation and curl is made more precise in the following section.

3.4.2 Relation between curl and rotation

Consider a rigid body rotating with angular velocity Ω. Then, as discussed in Section 1.3.1, the velocity $\boldsymbol{v}$ at any point can be written as $\boldsymbol{v} = \boldsymbol{\Omega} \times \boldsymbol{r}$, where the vector $\boldsymbol{\Omega}$ is directed along the axis of rotation (Figure 1.12).

Choosing the z-axis in the direction of $\boldsymbol{\Omega}$ gives $\boldsymbol{\Omega} = (0, 0, \Omega)$ and hence $\boldsymbol{v} = (0, 0, \Omega) \times (x, y, z) = (-\Omega y, \Omega x, 0)$. Computing the curl of this velocity field gives

$$\nabla \times \boldsymbol{v} = \left(-\frac{\partial(\Omega x)}{\partial z}, -\frac{\partial(\Omega y)}{\partial z}, \frac{\partial(\Omega x)}{\partial x} + \frac{\partial(\Omega y)}{\partial y}\right) = (0, 0, 2\Omega).$$

Hence $\nabla \times \boldsymbol{v} = 2\boldsymbol{\Omega}$, i.e. the curl of the velocity field is equal to twice the rotation rate.

3.4.3 Curl and conservative vector fields

Suppose that a vector field $\boldsymbol{u}$ is related to a scalar field ϕ by $\boldsymbol{u} = \nabla\phi$. Recall from Theorem 3.1 that this means that $\boldsymbol{u}$ is conservative. Now consider the curl of $\boldsymbol{u}$,

$$\begin{aligned} \nabla \times \boldsymbol{u} &= \boldsymbol{e}_1 \left(\frac{\partial}{\partial y}\left(\frac{\partial \phi}{\partial z}\right) - \frac{\partial}{\partial z}\left(\frac{\partial \phi}{\partial y}\right)\right) + \boldsymbol{e}_2 \left(\frac{\partial}{\partial z}\left(\frac{\partial \phi}{\partial x}\right) - \frac{\partial}{\partial x}\left(\frac{\partial \phi}{\partial z}\right)\right) \\ &\quad + \boldsymbol{e}_3 \left(\frac{\partial}{\partial x}\left(\frac{\partial \phi}{\partial y}\right) - \frac{\partial}{\partial y}\left(\frac{\partial \phi}{\partial x}\right)\right). \end{aligned}$$

Since the ordering of the cross derivatives of ϕ does not matter (see Section 3.1.1), all of these terms cancel out, giving the result that

$$\nabla \times \nabla\phi = \mathbf{0}. \tag{3.23}$$

Thus any vector field that can be written as the gradient of a scalar field is irrotational.

The converse of this result is also true, so that any irrotational vector field is conservative. This result will be proved in Section 5.2. In combination with Theorem 3.1, this means that the following three statements are equivalent:

- $\boldsymbol{u}$ can be written as the gradient of a potential: $\boldsymbol{u} = \boldsymbol{\nabla}\phi$.
- $\boldsymbol{u}$ is irrotational: $\boldsymbol{\nabla} \times \boldsymbol{u} = \boldsymbol{0}$.
- $\boldsymbol{u}$ is conservative: the line integral of $\boldsymbol{u}$ around any closed curve is zero.

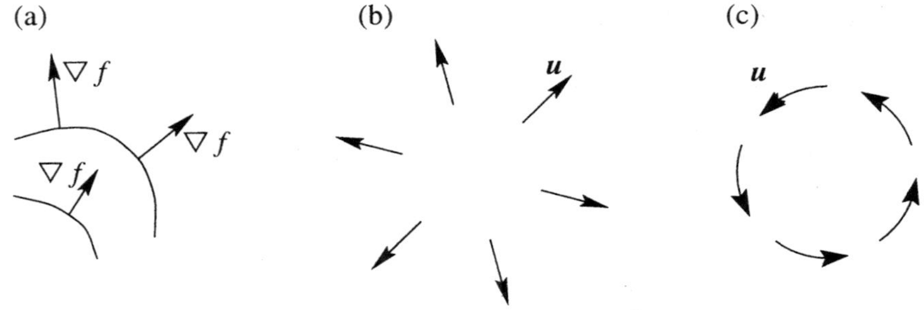

Fig. 3.8. Physical picture associated with (a) gradient, (b) divergence, (c) curl.

Summary of Chapter 3

- The *gradient* of a scalar field f, written grad f or ∇f, is a vector field perpendicular to the surfaces f = constant, pointing in the direction of increasing f, with magnitude equal to the rate of change of f in this direction.
- The components of ∇f are the partial derivatives of f:

$$\nabla f = \left(\frac{\partial f}{\partial x}, \frac{\partial f}{\partial y}, \frac{\partial f}{\partial z}\right).$$

- The *directional derivative* of f in the direction of the unit vector $\boldsymbol{u}$ is $\nabla f \cdot \boldsymbol{u}$.
- A vector field $\boldsymbol{F}$ is conservative if and only if it can be written as the gradient of a scalar field, $\boldsymbol{F} = \nabla\phi$. The function ϕ is called the *potential* for $\boldsymbol{F}$.
- The *divergence* of a vector field $\boldsymbol{u}$, written div $\boldsymbol{u}$ or $\nabla \cdot \boldsymbol{u}$, is a scalar field,

$$\nabla \cdot \boldsymbol{u} = \lim_{\delta V \to 0} \frac{1}{\delta V} \oiint_{\delta S} \boldsymbol{u} \cdot \boldsymbol{n}\, dS = \frac{\partial u_1}{\partial x} + \frac{\partial u_2}{\partial y} + \frac{\partial u_3}{\partial z}.$$

- The divergence of $\boldsymbol{u}$ corresponds to the amount of stretching or expansion associated with $\boldsymbol{u}$. If $\nabla \cdot \boldsymbol{u} = 0$, $\boldsymbol{u}$ is said to be *solenoidal.*
- The *Laplacian* of a scalar field ϕ, written $\nabla^2\phi$, is defined as $\nabla^2\phi = \nabla \cdot \nabla\phi$.
- The *curl* of a vector field $\boldsymbol{u}$ is a vector field, curl $\boldsymbol{u}$ or $\nabla \times \boldsymbol{u}$. Its component in the direction of a unit vector $\boldsymbol{n}$ perpendicular to the surface element δS is

$$\boldsymbol{n} \cdot \nabla \times \boldsymbol{u} = \lim_{\delta S \to 0} \frac{1}{\delta S} \oint_{\delta C} \boldsymbol{u} \cdot d\boldsymbol{r}.$$

- In component form,

$$\nabla \times \boldsymbol{u} = \left(\frac{\partial u_3}{\partial y} - \frac{\partial u_2}{\partial z}, \frac{\partial u_1}{\partial z} - \frac{\partial u_3}{\partial x}, \frac{\partial u_2}{\partial x} - \frac{\partial u_1}{\partial y}\right).$$

- Physically, $\nabla \times \boldsymbol{u}$ corresponds to the rotation or twist of $\boldsymbol{u}$. If $\nabla \times \boldsymbol{u} = \boldsymbol{0}$, $\boldsymbol{u}$ is said to be *irrotational.*
- If $\boldsymbol{u} = \nabla\phi$, then $\boldsymbol{u}$ is irrotational.
- The physical pictures corresponding to grad, div and curl are sketched in Figure 3.8.

EXERCISES

3.9 Find the gradient $\nabla\phi$ and the Laplacian $\nabla^2\phi$ for the scalar field $\phi = x^2 + xy + yz^2$.

3.10 Find the gradient and Laplacian of

$$\phi = \sin(kx)\sin(ly)\exp(\sqrt{k^2 + l^2}z).$$

3.11 Find the unit normal to the surface $xy^2 + 2yz = 4$ at the point $(-2, 2, 3)$.

3.12 For $\phi(x, y, z) = x^2 + y^2 + z^2 + xy - 3x$, find $\nabla\phi$ and find the minimum value of ϕ.

3.13 Find the equation of the plane which is tangent to the surface $x^2 + y^2 - 2z^3 = 0$ at the point $(1, 1, 1)$.

3.14 Find both the divergence and the curl of the vector fields
(a) $\boldsymbol{u} = (y, z, x)$;
(b) $\boldsymbol{v} = (xyz, z^2, x - y)$.

3.15 Show that both the divergence and the curl are linear operators, i.e. $\nabla \cdot (c\boldsymbol{u} + d\boldsymbol{v}) = c\nabla \cdot \boldsymbol{u} + d\nabla \cdot \boldsymbol{v}$ and $\nabla \times (c\boldsymbol{u} + d\boldsymbol{v}) = c\nabla \times \boldsymbol{u} + d\nabla \times \boldsymbol{v}$, where $\boldsymbol{u}$ and $\boldsymbol{v}$ are vector fields and c and d are constants.

3.16 For what values, if any, of the constants a and b is the vector field $\boldsymbol{u} = (y\cos x + axz, b\sin x + z, x^2 + y)$ irrotational?

3.17 (a) Show that $\boldsymbol{u} = (y^2z, -z^2\sin y + 2xyz, 2z\cos y + y^2x)$ is irrotational.
(b) Find the corresponding potential function.
(c) Hence find the value of the line integral of $\boldsymbol{u}$ along the curve $x = \sin\pi t/2$, $y = t^2 - t$, $z = t^4$, $0 \le t \le 1$.

4
Suffix Notation and its Applications

4.1 Introduction to suffix notation

This chapter introduces a powerful, compact notation for manipulating vector quantities. In the previous chapters, many of the vector expressions are awkward and cumbersome. This applies particularly to those expressions involving the cross product and the curl, such as the scalar triple product (1.8), the derivation of the alternative expression for the vector triple product (1.9) and the demonstration that $\boldsymbol{\nabla} \times \boldsymbol{\nabla}\phi = \mathbf{0}$ (3.23). Through the use of a new notation, *suffix notation*, such complicated expressions can be written much more concisely and many results can be proved more easily.

In this section, some simple vector equations are written using suffix notation. Consider first the equation $\boldsymbol{c} = \boldsymbol{a} + \boldsymbol{b}$. This vector equation is equivalent to the three equations for the components of $\boldsymbol{c}$, $c_i = a_i + b_i$ for $i = 1, 2, 3$. In suffix notation, the equation is simply written

$$c_i = a_i + b_i$$

and it is understood that this equation holds for $i = 1$, 2 and 3. The suffix i is called a 'free suffix'. The choice of this free suffix is arbitrary, so the equation could equally well be written $c_j = a_j + b_j$ or $c_k = a_k + b_k$. However, for simplicity and clarity the suffix i will be used for the free suffix in a vector equation in this book. Note that the same free suffix must be used for each term in the equation. This is an important rule which must be followed for an expression in suffix notation to be meaningful: the free suffix must match in each term in the expression.

Now consider the dot product of two vectors, $\boldsymbol{a} \cdot \boldsymbol{b} = a_1 b_1 + a_2 b_2 + a_3 b_3$. This can be written more compactly as

$$\boldsymbol{a} \cdot \boldsymbol{b} = \sum_{j=1}^{3} a_j b_j .$$

In suffix notation, this is written simply as

$$\boldsymbol{a} \cdot \boldsymbol{b} = a_j b_j , \tag{4.1}$$

where the repeated suffix j implies that the term is to be summed from $j = 1$ to $j = 3$. This is known as the *summation convention*: whenever a suffix is repeated in a single term in an equation, summation from 1 to 3 is understood. The repeated suffix is referred to as a 'dummy suffix', and must appear no more than twice in any term in an equation. The choice of the dummy suffix does not matter, so we can write $\boldsymbol{a} \cdot \boldsymbol{b} = a_j b_j = a_k b_k$, since each of these expressions is equivalent to $a_1 b_1 + a_2 b_2 + a_3 b_3$.

Next, suppose that an expression involves two dot products multiplied together, $(\boldsymbol{a} \cdot \boldsymbol{b})(\boldsymbol{c} \cdot \boldsymbol{d})$. In order to indicate which vector is dotted with which, a different dummy suffix must be used for each of the dot products:

$$(\boldsymbol{a} \cdot \boldsymbol{b})(\boldsymbol{c} \cdot \boldsymbol{d}) = a_j b_j c_k d_k .$$

Here, both j and k are repeated, so the summation convention implies summation over both j and k. Again the choice of dummy suffix is arbitrary, so for example we could have written $(\boldsymbol{a} \cdot \boldsymbol{b})(\boldsymbol{c} \cdot \boldsymbol{d}) = a_l b_l c_m d_m$. However, it is essential that no suffix appears more than twice in any term, since this would lead to ambiguity.

Note that it is the suffices that indicate which vector is dotted with which, not the ordering of the components of the vectors. In fact, since the components are just multiplied together, the ordering of terms is arbitrary, so the expression $c_k a_j d_k b_j$ also means $(\boldsymbol{a} \cdot \boldsymbol{b})(\boldsymbol{c} \cdot \boldsymbol{d})$.

Example 4.1

Write the suffix notation expression $a_j b_i c_j$ in ordinary vector notation.

The suffix j is repeated and is therefore a dummy suffix to be summed over. So $a_j b_i c_j$ means

$$\sum_{j=1}^{3} a_j b_i c_j = (\boldsymbol{a} \cdot \boldsymbol{c}) b_i ,$$

which is the i component of the vector $(\boldsymbol{a} \cdot \boldsymbol{c})\boldsymbol{b}$.

Example 4.2

Write the vector equation

$$\boldsymbol{u} + (\boldsymbol{a} \cdot \boldsymbol{b})\boldsymbol{v} = |\boldsymbol{a}|^2(\boldsymbol{b} \cdot \boldsymbol{v})\boldsymbol{a}$$

in suffix notation.

First introduce a free suffix i, which is understood to run from 1 to 3, to write the equation in component form; also write $|\boldsymbol{a}|^2$ as a dot product:

$$u_i + (\boldsymbol{a} \cdot \boldsymbol{b})v_i = (\boldsymbol{a} \cdot \boldsymbol{a})(\boldsymbol{b} \cdot \boldsymbol{v})a_i.$$

Now introduce a dummy suffix, which is repeated and therefore summed from 1 to 3, for each of the dot products:

$$u_i + a_j b_j v_i = a_j a_j b_k v_k a_i.$$

Note that two different dummy suffices are used on the right-hand side to avoid ambiguity.

Example 4.3

Show that the product of two $N \times N$ matrices A and B, $C = AB$ can be written in suffix notation as $C_{ij} = A_{ik}B_{kj}$. Hence show that the trace of the matrix AB (defined as the sum of the elements on the diagonal) is the same as the trace of BA.

C_{ij} is the element in the ith row and jth column of the matrix C. The rule for matrix multiplication is that if $C = AB$ then the element C_{ij} is obtained by taking the ith row of A and the jth column of B and multiplying these together term by term, so

$$C_{ij} = A_{i1}B_{1j} + A_{i2}B_{2j} \ldots + A_{iN}B_{Nj} = \sum_{k=1}^{N} A_{ik}B_{kj} = A_{ik}B_{kj},$$

where the repeated index k implies the sum from 1 to N.

The trace of the matrix C is the sum of the elements on the diagonal,

$$Tr(C) = C_{11} + C_{22} \ldots + C_{NN} = C_{jj}.$$

The trace of AB is

$$Tr(AB) = Tr(A_{ik}B_{kj}) = A_{jk}B_{kj}.$$

Similarly the trace of BA is

$$\begin{aligned} Tr(BA) &= B_{jk}A_{kj} \\ &= A_{kj}B_{jk} \quad \text{(since order of terms does not matter)} \\ &= A_{jk}B_{kj} \quad \text{(relabelling } j \leftrightarrow k) \\ &= Tr(AB). \end{aligned}$$

Note that this proof makes use of the fact that the choice of label used for a dummy suffix is arbitrary, so the labels j and k can be interchanged.

4.2 The Kronecker delta δ_{ij}

The *Kronecker delta* is written δ_{ij} and is defined by

$$\delta_{ij} = \begin{cases} 1 & \text{if } i = j, \\ 0 & \text{if } i \neq j. \end{cases} \tag{4.2}$$

The suffices i and j can each take the values 1, 2 or 3, so δ_{ij} has nine elements. From the above definition it follows that three of these are equal to 1 ($\delta_{11} = \delta_{22} = \delta_{33} = 1$) while the remaining six elements are equal to 0 ($\delta_{12} = \delta_{13} = \delta_{21} = \delta_{23} = \delta_{31} = \delta_{32} = 0$). δ_{ij} is an example of an object called a *tensor*. Tensors are described in detail in Chapter 7, but for the time being it is simplest to think of δ_{ij} as the 3×3 identity matrix,

$$\delta_{ij} = \begin{pmatrix} 1 & 0 & 0 \\ 0 & 1 & 0 \\ 0 & 0 & 1 \end{pmatrix}.$$

From the definition it is clear that δ_{ij} is symmetric, i.e. $\delta_{ij} = \delta_{ji}$.

Consider now the expression $\delta_{ij}a_j$. Notice that the suffix j is repeated, so by the summation convention, summation from $j = 1$ to 3 is understood. Hence

$$\delta_{ij}a_j = \sum_{j=1}^{3} \delta_{ij}a_j = \delta_{i1}a_1 + \delta_{i2}a_2 + \delta_{i3}a_3,$$

and it is clear that the result depends on the value of i. If $i = 1$, then $\delta_{i1} = 1$ while $\delta_{i2} = \delta_{i2} = 0$, so the right-hand side simplifies to a_1. Similarly, if $i = 2$ the result is a_2 and if $i = 3$ the result is a_3. In other words, the right-hand side simplifies to a_i, giving the important equation

$$\delta_{ij}a_j = a_i. \tag{4.3}$$

From the symmetry of δ_{ij} it follows that $\delta_{ij}a_i = a_j$. Because of this property of δ_{ij}, it is sometimes referred to as the 'substitution tensor', since its effect when multiplied by a_j is to replace the j with i.

There is a relationship between δ_{ij} and the dot product. The dot product of $\boldsymbol{a}$ and $\boldsymbol{b}$ can be written $\boldsymbol{a} \cdot \boldsymbol{b} = \delta_{ij}a_ib_j$. In this expression, both the i and the j suffices are repeated, so by the summation convention, both are to be summed from 1 to 3, giving a total of nine terms. However, because of the definition of δ_{ij}, only three of these terms (the ones with $i = j$) are non-zero, so

$$\delta_{ij}a_ib_j = \sum_{i=1}^{3}\sum_{j=1}^{3} \delta_{ij}a_ib_j = a_1b_1 + a_2b_2 + a_3b_3 = \boldsymbol{a} \cdot \boldsymbol{b}.$$

This result can also be demonstrated using (4.3), since $\delta_{ij}b_j = b_i$, so $\boldsymbol{a} \cdot \boldsymbol{b} = a_ib_i = a_i\delta_{ij}b_j = \delta_{ij}a_ib_j$.

Example 4.4

Evaluate δ_{jj}.

Since the suffix j is repeated, the summation convention implies that this expression must be summed from $j = 1$ to 3, so

$$\delta_{jj} = \sum_{j=1}^{3} \delta_{jj} = \delta_{11} + \delta_{22} + \delta_{33} = 3. \tag{4.4}$$

Example 4.5

Simplify $\delta_{ij}\delta_{jk}$.

Here the suffix j is repeated, and must therefore be summed over:

$$\delta_{ij}\delta_{jk} = \sum_{j=1}^{3} \delta_{ij}\delta_{jk} = \delta_{i1}\delta_{1k} + \delta_{i2}\delta_{2k} + \delta_{i3}\delta_{3k}.$$

The result depends on the values of i and k. If, for example, $i = 1$ and $k = 2$, we have $1\times0+0\times1+0\times0 = 0$; but if $i = 1$ and $k = 1$, we have $1\times1+0\times0+0\times0 = 1$. It is apparent that if i and k are different the result is 0, but if i and k are equal the result is 1. This result is therefore simply δ_{ik}, so the solution is $\delta_{ij}\delta_{jk} = \delta_{ik}$. Notice that this result is consistent with the substitution rule described above: the effect of δ_{ij} on δ_{jk} is to replace the j with i.

4.3 The alternating tensor ϵ_{ijk}

This section introduces the quantity which is used for writing cross products in suffix notation. This will prove extremely useful for manipulating expressions involving the cross product and the curl.

The *alternating tensor* is written ϵ_{ijk} and is defined by

$$\epsilon_{ijk} = \begin{cases} 0 & \text{if any of } i,\ j,\ k \text{ are equal,} \\ +1 & \text{if } (i,j,k) = (1,2,3),\ (2,3,1) \text{ or } (3,1,2), \\ -1 & \text{if } (i,j,k) = (1,3,2),\ (2,1,3) \text{ or } (3,2,1). \end{cases} \tag{4.5}$$

Since ϵ_{ijk} has three suffices, each of which can take any of the three values 1, 2 or 3, ϵ_{ijk} has 27 elements. However, from the above definition, all but six of these are zero. The six non-zero elements are $\epsilon_{123} = \epsilon_{231} = \epsilon_{312} = 1$ and $\epsilon_{132} = \epsilon_{213} = \epsilon_{321} = -1$.

There are two important symmetry properties of ϵ_{ijk} which follow directly from its definition:

- ϵ_{ijk} is unchanged if the suffices are reordered by moving them to the left and putting the first suffix third (a cyclic permutation of the suffices), i.e.

$$\epsilon_{ijk} = \epsilon_{jki} = \epsilon_{kij}. \tag{4.6}$$

- The sign of ϵ_{ijk} changes if any two of the suffices are interchanged, e.g.

$$\epsilon_{ijk} = -\epsilon_{jik}. \tag{4.7}$$

The relationship between ϵ_{ijk} and the cross product is as follows:

$$(\boldsymbol{a} \times \boldsymbol{b})_i = \epsilon_{ijk} a_j b_k. \tag{4.8}$$

In this equation, both j and k are repeated, so they are dummy suffices and must be summed over. To check that this agrees with the previous definition of $\boldsymbol{a} \times \boldsymbol{b}$, consider first the case $i = 1$. The right-hand side is then

$$\epsilon_{1jk} a_j b_k = \sum_{j=1}^{3} \sum_{k=1}^{3} \epsilon_{1jk} a_j b_k.$$

Since ϵ_{ijk} is only non-zero when all three of its suffices are different, only the two terms $j = 2$, $k = 3$ and $j = 3$, $k = 2$ are non-zero in the double sum. Hence the right-hand side reduces to $\epsilon_{123} a_2 b_3 + \epsilon_{132} a_3 b_2 = a_2 b_3 - a_3 b_2$. This agrees with the previous definition (1.4) for the first component of the cross product. It can be seen by cyclic permutation of indices that the second and third components also agree.

There is also a relation between ϵ_{ijk} and the determinant of a 3×3 matrix. This can be written

$$|M| = \epsilon_{ijk} M_{1i} M_{2j} M_{3k}. \tag{4.9}$$

A related formula is

$$\epsilon_{pqr}|M| = \epsilon_{ijk} M_{pi} M_{qj} M_{rk}. \tag{4.10}$$

An expression for the scalar triple product $\boldsymbol{a} \cdot \boldsymbol{b} \times \boldsymbol{c}$ can be deduced in suffix notation as follows:

$$\boldsymbol{a} \cdot \boldsymbol{b} \times \boldsymbol{c} = a_i(\boldsymbol{b} \times \boldsymbol{c})_i = a_i \epsilon_{ijk} b_j c_k = \epsilon_{ijk} a_i b_j c_k. \tag{4.11}$$

A comparison of this neat and elegant expression with the cumbersome formula in terms of the components (1.8) shows the power of suffix notation. The properties of the scalar triple product can also be deduced using suffix notation, as in the following examples.

Example 4.6

Use suffix notation to show that $\boldsymbol{a} \cdot \boldsymbol{b} \times \boldsymbol{c} = \boldsymbol{a} \times \boldsymbol{b} \cdot \boldsymbol{c}$.

$$\begin{aligned}
\boldsymbol{a} \cdot \boldsymbol{b} \times \boldsymbol{c} &= \epsilon_{ijk} a_i b_j c_k \\
&= \epsilon_{kij} a_i b_j c_k \quad (\text{using } \epsilon_{ijk} = \epsilon_{kij}) \\
&= (\boldsymbol{a} \times \boldsymbol{b})_k c_k \\
&= \boldsymbol{a} \times \boldsymbol{b} \cdot \boldsymbol{c}.
\end{aligned}$$

Example 4.7

Show that $\boldsymbol{a} \cdot \boldsymbol{b} \times \boldsymbol{c} = \boldsymbol{b} \cdot \boldsymbol{c} \times \boldsymbol{a}$.

The demonstration of this result is very similar:

$$\begin{aligned}
\boldsymbol{a} \cdot \boldsymbol{b} \times \boldsymbol{c} &= \epsilon_{ijk} a_i b_j c_k \\
&= \epsilon_{jki} a_i b_j c_k \quad (\text{using } \epsilon_{ijk} = \epsilon_{jki}) \\
&= b_j \epsilon_{jki} c_k a_i \quad (\text{just rearranging terms}) \\
&= b_j (\boldsymbol{c} \times \boldsymbol{a})_j \\
&= \boldsymbol{b} \cdot \boldsymbol{c} \times \boldsymbol{a}.
\end{aligned}$$

Example 4.8

Evaluate ϵ_{iik}.

Since $\epsilon_{ijk} = 0$ if any of i, j, k are equal, it follows that $\epsilon_{iik} = 0$.

Example 4.9

Evaluate $\epsilon_{ijk}\epsilon_{ijk}$.

In this expression all three suffices i, j and k are repeated, and must therefore be summed over, giving a total of 27 terms. Only six of these terms are non-zero, so $\epsilon_{ijk}\epsilon_{ijk} = \epsilon_{123}^2 + \epsilon_{132}^2 + \epsilon_{213}^2 + \epsilon_{231}^2 + \epsilon_{312}^2 + \epsilon_{321}^2 = 6$.

4.4 Relation between ϵ_{ijk} and δ_{ij}

An important relationship between ϵ_{ijk} and δ_{ij} is the following equation:

$$\epsilon_{ijk}\epsilon_{klm} = \delta_{il}\delta_{jm} - \delta_{im}\delta_{jl}. \tag{4.12}$$

This equation has four free suffices (i, j, l and m) and therefore represents 81 different equations! The left-hand side is summed over k, because the suffix k appears twice.

The result (4.12) can be demonstrated by the following argument: since the three coordinate axes are equivalent, we need only consider the case $i = 1$. Consider now the possible values for j:

1. If $j = 1$, $\epsilon_{ijk} = \epsilon_{11k} = 0$ and so the l.h.s. is zero; the r.h.s. is $\delta_{1l}\delta_{1m} - \delta_{1m}\delta_{1l}$ which is also zero since the two δ terms cancel.
2. If $j = 2$, $\epsilon_{ijk} = \epsilon_{12k} = 0$ unless $k = 3$, so only the $k = 3$ term contributes to the sum. When $k = 3$, the term ϵ_{klm} is zero unless l and m are 1 and 2. Therefore the l.h.s. takes the value +1 if $l = 1$ and $m = 2$, -1 if $l = 2$ and $m = 1$, and zero otherwise. Now the r.h.s. is $\delta_{1l}\delta_{2m} - \delta_{1m}\delta_{2l}$. This is also equal to +1 when $l = 1$ and $m = 2$ (from the first term), -1 when $l = 2$ and $m = 1$ (from the second term) and zero otherwise.
3. If $j = 3$, an equivalent argument to the case $j = 2$ applies; the details are left to the reader.

Equation (4.12) is very useful for simplifying expressions involving two cross products.

Example 4.10

Derive the formula (1.9) for the expansion of the vector triple product using suffix notation.

$$(\boldsymbol{a} \times (\boldsymbol{b} \times \boldsymbol{c}))_i = \epsilon_{ijk} a_j (\boldsymbol{b} \times \boldsymbol{c})_k$$

(writing the first cross product in suffix notation)

$$= \epsilon_{ijk} a_j \epsilon_{klm} b_l c_m$$

(writing the second cross product in suffix notation)

$$= (\delta_{il}\delta_{jm} - \delta_{im}\delta_{jl}) a_j b_l c_m \quad \text{(using (4.12))}$$

$$= a_m b_i c_m - a_j b_j c_i \quad \text{(using (4.3))}$$

$$= (\boldsymbol{a} \cdot \boldsymbol{c}) b_i - (\boldsymbol{a} \cdot \boldsymbol{b}) c_i,$$

so we have shown that

$$\boldsymbol{a} \times (\boldsymbol{b} \times \boldsymbol{c}) = (\boldsymbol{a} \cdot \boldsymbol{c})\boldsymbol{b} - (\boldsymbol{a} \cdot \boldsymbol{b})\boldsymbol{c}. \tag{4.13}$$

Notice one very important point in this analysis: in the second line, the k component of the vector $\boldsymbol{b} \times \boldsymbol{c}$ is required. When this is written down in suffix notation it is essential that 'new' suffices are used (l and m above) to avoid repeating the existing suffices i and j – recall the essential rule of suffix notation that no suffix may appear more than twice.

EXERCISES

4.1 Write the vector equation $\boldsymbol{a} \times \boldsymbol{b} + (\boldsymbol{a} \cdot \boldsymbol{d})\boldsymbol{c} = \boldsymbol{e}$ in suffix notation.

4.2 Translate the suffix notation equation $\delta_{ij} c_j + \epsilon_{kji} a_k b_j = d_l e_m c_i b_l c_m$ into ordinary vector notation.

4.3 Use suffix notation to show that $\boldsymbol{a} \times \boldsymbol{b} = -\boldsymbol{b} \times \boldsymbol{a}$.

4.4 Simplify the suffix notation expressions

(a) $\delta_{ij}\epsilon_{ijk}$;

(b) $\epsilon_{ijk}\epsilon_{ilm}$;

(c) $\epsilon_{ijk}\epsilon_{ijm}$;

(d) $\epsilon_{ijk}\epsilon_{ijk}$.

4.5 Using suffix notation, find an alternative expression (involving no cross products) for $\boldsymbol{a} \times \boldsymbol{b} \cdot \boldsymbol{c} \times \boldsymbol{d}$.

4.6 If A and B are two $N \times N$ matrices, show that $(AB)^T = B^T A^T$, where A^T is the *transpose* of A defined by interchanging the rows and columns of A.

4.7 Verify the formulae (4.9) and (4.10) for the determinant of a 3×3 matrix.

4.8 Use the formula (4.10) for the determinant of a 3×3 matrix M to show that

(a) $6|M| = \epsilon_{pqr}\epsilon_{ijk} M_{pi} M_{qj} M_{rk}$;

(b) $|M^T| = |M|$;

(c) $|MN| = |M||N|$.

4.5 Grad, div and curl in suffix notation

The differential operators grad, div and curl can be written using suffix notation. To do this, the Cartesian coordinates (x, y, z) will be relabelled (x_1, x_2, x_3). As in the previous section, the use of suffix notation results in a much more compact formulation and simplifies many of the computations.

Consider first the gradient of a scalar field, $\boldsymbol{\nabla} f$. This is defined by

$$\boldsymbol{\nabla} f = \left(\frac{\partial f}{\partial x_1}, \frac{\partial f}{\partial x_2}, \frac{\partial f}{\partial x_3} \right).$$

The i component of $\boldsymbol{\nabla} f$ is equal to the partial derivative of f with respect to x_i, so in suffix notation this can be written

$$[\boldsymbol{\nabla} f]_i = \frac{\partial f}{\partial x_i}. \tag{4.14}$$

Thus the vector differential operator $\boldsymbol{\nabla}$ defined in (3.12) can be written in suffix notation as

$$[\boldsymbol{\nabla}]_i = \frac{\partial}{\partial x_i}. \tag{4.15}$$

The divergence of a vector field $\boldsymbol{u}$ is

$$\boldsymbol{\nabla} \cdot \boldsymbol{u} = \frac{\partial u_1}{\partial x_1} + \frac{\partial u_2}{\partial x_2} + \frac{\partial u_3}{\partial x_3} = \frac{\partial u_j}{\partial x_j}, \tag{4.16}$$

where the summation convention implies the sum over j from 1 to 3. Note that the same expression results from taking the dot product of $\boldsymbol{\nabla}$ defined by (4.15) with the vector $\boldsymbol{u}$ using the suffix notation formula (4.1) for the dot product of two vectors.

The first component of $\boldsymbol{\nabla} \times \boldsymbol{u}$ is, from (3.22),

$$[\boldsymbol{\nabla} \times \boldsymbol{u}]_1 = \frac{\partial u_3}{\partial x_2} - \frac{\partial u_2}{\partial x_3} = \epsilon_{1jk} \frac{\partial u_k}{\partial x_j},$$

where the repeated j and k imply a double sum, and so the suffix notation expression for $\boldsymbol{\nabla} \times \boldsymbol{u}$ is

$$[\boldsymbol{\nabla} \times \boldsymbol{u}]_i = \epsilon_{ijk} \frac{\partial u_k}{\partial x_j}. \tag{4.17}$$

This can also be obtained simply by taking the cross product of $\boldsymbol{\nabla}$ with $\boldsymbol{u}$ using (4.8).

Example 4.11

Let $\boldsymbol{r}$ be the position vector $\boldsymbol{r} = (x_1, x_2, x_3)$ and $r = |\boldsymbol{r}|$. Use suffix notation to evaluate $\partial x_i/\partial x_j$. Hence find $\boldsymbol{\nabla} r$, $\boldsymbol{\nabla} \cdot \boldsymbol{r}$ and $\boldsymbol{\nabla} \times \boldsymbol{r}$.

$\boldsymbol{r} = (x_1, x_2, x_3)$, so in suffix notation, $r_i = x_i$. The three coordinate axes x_1, x_2, x_3 are independent. Thus the derivative of each of the x_i with respect to one of the others is zero, while the derivative with respect to itself is 1. Thus

$$\begin{aligned} \frac{\partial x_i}{\partial x_j} &= \begin{cases} 1 & \text{if } i = j, \\ 0 & \text{if } i \neq j \end{cases} \\ &= \delta_{ij}. \end{aligned} \tag{4.18}$$

To find $\boldsymbol{\nabla} r$, first write $r = |\boldsymbol{r}| = (\boldsymbol{r} \cdot \boldsymbol{r})^{1/2} = (x_j x_j)^{1/2}$, so

$$\begin{aligned} [\boldsymbol{\nabla} r]_i &= \frac{\partial}{\partial x_i}(x_j x_j)^{1/2} \\ &= \frac{1}{2}(x_j x_j)^{-1/2}\frac{\partial}{\partial x_i}(x_j x_j) \\ &= \frac{1}{2r}2x_j\frac{\partial x_j}{\partial x_i} \\ &= \frac{1}{r}x_j\delta_{ij} = \frac{x_i}{r}. \end{aligned}$$

So

$$\boldsymbol{\nabla} r = \boldsymbol{r}/r \tag{4.19}$$

as was shown in Exercise 3.3 without the use of suffix notation. Similarly, the divergence of $\boldsymbol{r}$ is

$$\boldsymbol{\nabla} \cdot \boldsymbol{r} = \frac{\partial x_j}{\partial x_j} = \delta_{jj} = 3$$

and the curl of $\boldsymbol{r}$ is

$$[\boldsymbol{\nabla} \times \boldsymbol{r}]_i = \epsilon_{ijk}\frac{\partial x_k}{\partial x_j} = \epsilon_{ijk}\delta_{jk} = 0,$$

using the result of Exercise 4.4(a).

4.6 Combinations of grad, div and curl

The operators grad, div and curl can be combined together in several different ways. Some of these combinations can be simplified or expanded into alternative expressions. These combinations are considered below, making use of suffix notation. All of these results can also be obtained without suffix notation, by writing out all the components, but in most cases the suffix notation method is much quicker.

- Div grad:
$$\boldsymbol{\nabla} \cdot (\boldsymbol{\nabla} f) = \frac{\partial}{\partial x_j}\left([\boldsymbol{\nabla} f]_j\right) = \frac{\partial}{\partial x_j}\left(\frac{\partial f}{\partial x_j}\right) = \frac{\partial^2 f}{\partial x_j \partial x_j} = \nabla^2 f. \tag{4.20}$$
This is the Laplacian of f introduced in Section 3.3.2, the sum of the second partial derivatives of f.
- Curl grad: this combination was shown to be zero in Section 3.4.3. This result can be shown using suffix notation as follows:
$$\begin{aligned} [\boldsymbol{\nabla} \times (\boldsymbol{\nabla} f)]_i &= \epsilon_{ijk}\frac{\partial}{\partial x_j}\frac{\partial f}{\partial x_k} \\ &= \epsilon_{ikj}\frac{\partial}{\partial x_k}\frac{\partial f}{\partial x_j} \quad \text{(relabelling } j \leftrightarrow k\text{)} \\ &= -\epsilon_{ijk}\frac{\partial}{\partial x_k}\frac{\partial f}{\partial x_j} \quad \text{(using } \epsilon_{ikj} = -\epsilon_{ijk}\text{)} \\ &= -\epsilon_{ijk}\frac{\partial}{\partial x_j}\frac{\partial f}{\partial x_k} \quad \text{(as order of derivatives does not matter)} \\ &= 0, \end{aligned}$$
since the expression has been manipulated to give minus itself.
- Grad div:
$$[\boldsymbol{\nabla}(\boldsymbol{\nabla} \cdot \boldsymbol{u})]_i = \frac{\partial}{\partial x_i}\left(\frac{\partial u_j}{\partial x_j}\right) = \frac{\partial^2 u_j}{\partial x_i \partial x_j}. \tag{4.21}$$
This quantity cannot be simplified further.
- Div curl:
$$\begin{aligned} \boldsymbol{\nabla} \cdot (\boldsymbol{\nabla} \times \boldsymbol{u}) &= \frac{\partial}{\partial x_i}\epsilon_{ijk}\frac{\partial u_k}{\partial x_j} \\ &= \epsilon_{jik}\frac{\partial}{\partial x_j}\frac{\partial u_k}{\partial x_i} \quad \text{(relabelling } i \leftrightarrow j\text{)} \\ &= -\epsilon_{ijk}\frac{\partial}{\partial x_j}\frac{\partial u_k}{\partial x_i} \end{aligned}$$

$$
\begin{aligned}
&= -\epsilon_{ijk}\frac{\partial}{\partial x_i}\frac{\partial u_k}{\partial x_j} \\
&= 0, \qquad (4.22)
\end{aligned}
$$

using exactly the same argument as for curl grad.

- Curl curl:

$$
\begin{aligned}
[\nabla \times (\nabla \times \boldsymbol{u})]_i &= \epsilon_{ijk}\frac{\partial}{\partial x_j}\epsilon_{klm}\frac{\partial u_m}{\partial x_l} \\
&= \epsilon_{ijk}\epsilon_{klm}\frac{\partial^2 u_m}{\partial x_j \partial x_l} \\
&= (\delta_{il}\delta_{jm} - \delta_{im}\delta_{jl})\frac{\partial^2 u_m}{\partial x_j \partial x_l} \\
&= \frac{\partial^2 u_j}{\partial x_j \partial x_i} - \frac{\partial^2 u_i}{\partial x_j \partial x_j} \\
&= \left[\nabla(\nabla \cdot \boldsymbol{u}) - \nabla^2 \boldsymbol{u}\right]_i . \qquad (4.23)
\end{aligned}
$$

This result can be used to provide a physical definition for ∇^2 applied to a vector. The previous definition (3.19) was only defined in terms of the components of the vector in Cartesian coordinates. From the above result, $\nabla^2 \boldsymbol{u}$ can be defined by

$$
\nabla^2 \boldsymbol{u} = \nabla(\nabla \cdot \boldsymbol{u}) - \nabla \times (\nabla \times \boldsymbol{u}). \qquad (4.24)
$$

These five combinations of grad, div and curl are the only ones that make sense. For example, the combination grad curl has no meaning since curl is a vector but grad can only act on a scalar. Combinations of three or more of the operators grad, div and curl can be evaluated using the above results, as in the following example.

Example 4.12

Show that

$$
\nabla \times (\nabla^2 \boldsymbol{u}) = \nabla^2(\nabla \times \boldsymbol{u}). \qquad (4.25)
$$

Using the result (4.24),

$$
\begin{aligned}
\nabla \times (\nabla^2 \boldsymbol{u}) &= \nabla \times \big(\nabla(\nabla \cdot \boldsymbol{u}) - \nabla \times (\nabla \times \boldsymbol{u})\big) \\
&= -\nabla \times \big(\nabla \times (\nabla \times \boldsymbol{u})\big) \quad (\text{since } \nabla \times \nabla = 0) \\
&= -\nabla\big(\nabla \cdot (\nabla \times \boldsymbol{u})\big) + \nabla^2(\nabla \times \boldsymbol{u}) \quad (\text{using (4.23)}) \\
&= \nabla^2(\nabla \times \boldsymbol{u}) \quad (\text{since } \nabla \cdot \nabla \times \boldsymbol{u} = 0).
\end{aligned}
$$

So the operators ∇^2 and $\nabla\times$ commute.

4.7 Grad, div and curl applied to products of functions

Another useful application of suffix notation is in computing the action of grad, div and curl on products of vector and scalar fields. As in the previous section, these results can also be obtained by writing out all the components, but the suffix notation method is much more compact and elegant. Some of these results are straightforward applications of the usual rule for the differentiation of a product and can simply be written down without any calculation, but many of them are not so obvious.

In the following, f and g are differentiable scalar fields and $\boldsymbol{u}$ and $\boldsymbol{v}$ are differentiable vector fields.

$$[\boldsymbol{\nabla}(fg)]_i = \frac{\partial}{\partial x_i}(fg) = f\frac{\partial g}{\partial x_i} + g\frac{\partial f}{\partial x_i} = [f\boldsymbol{\nabla} g + g\boldsymbol{\nabla} f]_i, \quad \text{so}$$

$$\boldsymbol{\nabla}(fg) = f\boldsymbol{\nabla} g + g\boldsymbol{\nabla} f. \tag{4.26}$$

$$\begin{aligned}\boldsymbol{\nabla}\cdot(f\boldsymbol{u}) &= \frac{\partial}{\partial x_i}(fu_i)\\ &= \frac{\partial f}{\partial x_i}u_i + f\frac{\partial u_i}{\partial x_i}\\ &= \boldsymbol{\nabla} f\cdot\boldsymbol{u} + f\boldsymbol{\nabla}\cdot\boldsymbol{u}.\end{aligned} \tag{4.27}$$

$$\begin{aligned}[\boldsymbol{\nabla}\times(f\boldsymbol{u})]_i &= \epsilon_{ijk}\frac{\partial}{\partial x_j}(f\boldsymbol{u}_k)\\ &= \epsilon_{ijk}\frac{\partial f}{\partial x_j}\boldsymbol{u}_k + f\epsilon_{ijk}\frac{\partial u_k}{\partial x_j}\\ &= [\boldsymbol{\nabla} f\times\boldsymbol{u} + f\boldsymbol{\nabla}\times\boldsymbol{u}]_i.\end{aligned} \tag{4.28}$$

$$\begin{aligned}\boldsymbol{\nabla}\cdot(\boldsymbol{u}\times\boldsymbol{v}) &= \frac{\partial}{\partial x_i}(\epsilon_{ijk}u_j v_k)\\ &= \epsilon_{ijk}\frac{\partial u_j}{\partial x_i}v_k + \epsilon_{ijk}u_j\frac{\partial v_k}{\partial x_i}\\ &= \left(\epsilon_{kij}\frac{\partial u_j}{\partial x_i}\right)v_k - \left(\epsilon_{jik}\frac{\partial v_k}{\partial x_i}\right)u_j\\ &= (\boldsymbol{\nabla}\times\boldsymbol{u})\cdot\boldsymbol{v} - (\boldsymbol{\nabla}\times\boldsymbol{v})\cdot\boldsymbol{u}.\end{aligned} \tag{4.29}$$

$$\begin{aligned}
[\nabla \times (\boldsymbol{u} \times \boldsymbol{v})]_i &= \epsilon_{ijk} \frac{\partial}{\partial x_j}(\epsilon_{klm} u_l v_m) \\
&= (\delta_{il}\delta_{jm} - \delta_{im}\delta_{jl}) \frac{\partial}{\partial x_j}(u_l v_m) \\
&= \frac{\partial}{\partial x_j}(u_i v_j) - \frac{\partial}{\partial x_j}(u_j v_i) \\
&= u_i \frac{\partial v_j}{\partial x_j} + v_j \frac{\partial u_i}{\partial x_j} - u_j \frac{\partial v_i}{\partial x_j} - v_i \frac{\partial u_j}{\partial x_j} \\
&= [\boldsymbol{u}(\nabla \cdot \boldsymbol{v}) + \boldsymbol{v} \cdot \nabla \boldsymbol{u} - \boldsymbol{u} \cdot \nabla \boldsymbol{v} - \boldsymbol{v}(\nabla \cdot \boldsymbol{u})]_i, \quad (4.30)
\end{aligned}$$

where the operator $\boldsymbol{u} \cdot \nabla$ is defined by

$$\boldsymbol{u} \cdot \nabla = u_j \frac{\partial}{\partial x_j} \quad (4.31)$$

and can act on either a scalar or a vector.

To find an expansion for the expression $\nabla(\boldsymbol{u} \cdot \boldsymbol{v})$, consider first the quantity

$$\begin{aligned}
[\boldsymbol{u} \times (\nabla \times \boldsymbol{v})]_i &= \epsilon_{ijk} u_j \epsilon_{klm} \frac{\partial v_m}{\partial x_l} \\
&= (\delta_{il}\delta_{jm} - \delta_{im}\delta_{jl}) u_j \frac{\partial v_m}{\partial x_l} \\
&= u_j \frac{\partial v_j}{\partial x_i} - u_j \frac{\partial v_i}{\partial x_j}.
\end{aligned}$$

Similarly, interchanging $\boldsymbol{u}$ and $\boldsymbol{v}$,

$$[\boldsymbol{v} \times (\nabla \times \boldsymbol{u})]_i = v_j \frac{\partial u_j}{\partial x_i} - v_j \frac{\partial u_i}{\partial x_j}.$$

Adding these two equations gives

$$\begin{aligned}
[\boldsymbol{u} \times (\nabla \times \boldsymbol{v}) + \boldsymbol{v} \times (\nabla \times \boldsymbol{u})]_i &= u_j \frac{\partial v_j}{\partial x_i} - u_j \frac{\partial v_i}{\partial x_j} + v_j \frac{\partial u_j}{\partial x_i} - v_j \frac{\partial u_i}{\partial x_j} \\
&= [\nabla(\boldsymbol{u} \cdot \boldsymbol{v}) - \boldsymbol{u} \cdot \nabla \boldsymbol{v} - \boldsymbol{v} \cdot \nabla \boldsymbol{u}]_i. \quad (4.32)
\end{aligned}$$

This can be rearranged to give

$$\nabla(\boldsymbol{u} \cdot \boldsymbol{v}) = \boldsymbol{u} \times (\nabla \times \boldsymbol{v}) + \boldsymbol{v} \times (\nabla \times \boldsymbol{u}) + \boldsymbol{u} \cdot \nabla \boldsymbol{v} + \boldsymbol{v} \cdot \nabla \boldsymbol{u}. \quad (4.33)$$

The effect of applying grad, div or curl to products of more than two scalar or vector functions can be obtained either by the repeated use of the above results, or directly by suffix notation, as in the following example.

Example 4.13

Find an expansion for $\boldsymbol{\nabla}\cdot(fg\boldsymbol{u})$.

$$\begin{aligned}\boldsymbol{\nabla}\cdot(fg\boldsymbol{u}) &= \boldsymbol{\nabla}(fg)\cdot\boldsymbol{u}+(fg)\boldsymbol{\nabla}\cdot\boldsymbol{u} \quad \text{(using (4.27))}\\ &= (f\boldsymbol{\nabla}g+g\boldsymbol{\nabla}f)\cdot\boldsymbol{u}+(fg)\boldsymbol{\nabla}\cdot\boldsymbol{u}\\ &= f\boldsymbol{\nabla}g\cdot\boldsymbol{u}+g\boldsymbol{\nabla}f\cdot\boldsymbol{u}+fg\boldsymbol{\nabla}\cdot\boldsymbol{u}.\end{aligned}$$

Alternatively, using suffix notation,

$$\begin{aligned}\boldsymbol{\nabla}\cdot(fg\boldsymbol{u}) &= \frac{\partial}{\partial x_i}(fgu_i)\\ &= fg\frac{\partial u_i}{\partial x_i}+f\frac{\partial g}{\partial x_i}u_i+\frac{\partial f}{\partial x_i}gu_i\\ &= fg\boldsymbol{\nabla}\cdot\boldsymbol{u}+f\boldsymbol{\nabla}g\cdot\boldsymbol{u}+g\boldsymbol{\nabla}f\cdot\boldsymbol{u}.\end{aligned}$$

Example 4.14

Show that $\boldsymbol{u}\cdot\boldsymbol{\nabla}\boldsymbol{u}=\boldsymbol{\nabla}(|\boldsymbol{u}|^2/2)-\boldsymbol{u}\times(\boldsymbol{\nabla}\times\boldsymbol{u})$.

Apply (4.33) with $\boldsymbol{v}=\boldsymbol{u}$:

$$\boldsymbol{\nabla}(\boldsymbol{u}\cdot\boldsymbol{u})=2\boldsymbol{u}\times(\boldsymbol{\nabla}\times\boldsymbol{u})+2\boldsymbol{u}\cdot\boldsymbol{\nabla}\boldsymbol{u}.$$

Rearranging this and dividing by 2 gives

$$\boldsymbol{u}\cdot\boldsymbol{\nabla}\boldsymbol{u}=\boldsymbol{\nabla}(|\boldsymbol{u}|^2/2)-\boldsymbol{u}\times(\boldsymbol{\nabla}\times\boldsymbol{u}). \tag{4.34}$$

Example 4.15

Use the results (4.30) and (4.33) to provide a definition of $\boldsymbol{u}\cdot\boldsymbol{\nabla}\boldsymbol{v}$ that is not given in terms of Cartesian components.

By subtracting (4.30) from (4.33), $\boldsymbol{v}\cdot\boldsymbol{\nabla}\boldsymbol{u}$ is eliminated and we obtain

$$\boldsymbol{\nabla}(\boldsymbol{u}\cdot\boldsymbol{v})-\boldsymbol{\nabla}\times(\boldsymbol{u}\times\boldsymbol{v})=\boldsymbol{u}\times(\boldsymbol{\nabla}\times\boldsymbol{v})+\boldsymbol{v}\times(\boldsymbol{\nabla}\times\boldsymbol{u})+2\boldsymbol{u}\cdot\boldsymbol{\nabla}\boldsymbol{v}-\boldsymbol{u}(\boldsymbol{\nabla}\cdot\boldsymbol{v})+\boldsymbol{v}(\boldsymbol{\nabla}\cdot\boldsymbol{u})$$

which can be rearranged to give

$$\begin{aligned}\boldsymbol{u}\cdot\boldsymbol{\nabla}\boldsymbol{v} &= \frac{1}{2}\big(\boldsymbol{\nabla}(\boldsymbol{u}\cdot\boldsymbol{v})-\boldsymbol{\nabla}\times(\boldsymbol{u}\times\boldsymbol{v})-\boldsymbol{u}\times(\boldsymbol{\nabla}\times\boldsymbol{v})-\boldsymbol{v}\times(\boldsymbol{\nabla}\times\boldsymbol{u})\\ &\quad +\boldsymbol{u}(\boldsymbol{\nabla}\cdot\boldsymbol{v})-\boldsymbol{v}(\boldsymbol{\nabla}\cdot\boldsymbol{u})\big).\end{aligned} \tag{4.35}$$

Summary of Chapter 4

Suffix notation

Suffix notation is a powerful tool for manipulating expressions involving vectors. The rules of suffix notation are as follows:

- Within any term in an equation, any suffix must appear either once or twice. No suffix may appear more that twice.
- A suffix that appears once in any term is called a 'free' suffix. A free suffix takes the values 1, 2 and 3 and represents the components of a vector. For example $\boldsymbol{a} + \boldsymbol{b} = \boldsymbol{c} - \boldsymbol{d}$ is written in suffix notation as $a_i + b_i = c_i - d_i$.
- In a vector equation, the free suffix must be the same in each term. The above equation may also be written $a_j + b_j = c_j - d_j$, or any other suffix may be used, provided the same suffix appears in each term.
- A suffix that appears twice in a term is called a 'dummy' suffix and is summed from 1 to 3. This is known as the *summation convention.* So $a_j b_j$ means $a_1 b_1 + a_2 b_2 + a_3 b_3 = \boldsymbol{a} \cdot \boldsymbol{b}$.
- A pair of dummy suffices can be changed. For example, $a_j b_j$, $a_k b_k$ and $a_m b_m$ are all equal to $\boldsymbol{a} \cdot \boldsymbol{b}$.
- The order of terms in a suffix notation expression does not matter.
- The *Kronecker delta* is defined by $\delta_{ij} = 1$ if $i = j$, 0 if $i \neq j$. Properties include $\delta_{ij} = \delta_{ji}$, $\delta_{ij} a_j = a_i$ and $\delta_{ij} a_i b_j = \boldsymbol{a} \cdot \boldsymbol{b}$.
- The *alternating tensor* ϵ_{ijk} is defined by $\epsilon_{ijk} = 0$ if any of i, j, k are equal, $\epsilon_{123} = \epsilon_{231} = \epsilon_{312} = 1$, $\epsilon_{132} = \epsilon_{213} = \epsilon_{321} = -1$. Properties include $\epsilon_{ijk} = \epsilon_{jki} = \epsilon_{kij}$, $\epsilon_{ijk} = -\epsilon_{jik}$.
- The cross product of $\boldsymbol{a}$ and $\boldsymbol{b}$ can be written $[\boldsymbol{a} \times \boldsymbol{b}]_i = \epsilon_{ijk} a_j b_k$.
- ϵ_{ijk} and δ_{ij} are related by $\epsilon_{ijk}\epsilon_{klm} = \delta_{il}\delta_{jm} - \delta_{im}\delta_{jl}$.
- Grad, div and curl can be written in suffix notation as follows:

$$[\boldsymbol{\nabla} f]_i = \frac{\partial f}{\partial x_i}, \qquad \boldsymbol{\nabla} \cdot \boldsymbol{u} = \frac{\partial u_j}{\partial x_j}, \qquad [\boldsymbol{\nabla} \times \boldsymbol{u}]_i = \epsilon_{ijk} \frac{\partial u_k}{\partial x_j}.$$

Combinations of operators and derivatives of products

- $\boldsymbol{\nabla} \cdot (\boldsymbol{\nabla} f) = \nabla^2 f$.
- $\boldsymbol{\nabla} \times (\boldsymbol{\nabla} f) = \boldsymbol{0}$.
- $\boldsymbol{\nabla} \cdot (\boldsymbol{\nabla} \times \boldsymbol{u}) = 0$.
- $\boldsymbol{\nabla} \times (\boldsymbol{\nabla} \times \boldsymbol{u}) = \boldsymbol{\nabla}(\boldsymbol{\nabla} \cdot \boldsymbol{u}) - \nabla^2 \boldsymbol{u}$.

- $\boldsymbol{\nabla}(fg) = f\boldsymbol{\nabla} g + g\boldsymbol{\nabla} f$.
- $\boldsymbol{\nabla}\cdot(f\boldsymbol{u}) = \boldsymbol{\nabla} f\cdot\boldsymbol{u} + f\boldsymbol{\nabla}\cdot\boldsymbol{u}$.
- $\boldsymbol{\nabla}\times(f\boldsymbol{u}) = \boldsymbol{\nabla} f\times\boldsymbol{u} + f\boldsymbol{\nabla}\times\boldsymbol{u}$.
- $\boldsymbol{\nabla}\cdot(\boldsymbol{u}\times\boldsymbol{v}) = (\boldsymbol{\nabla}\times\boldsymbol{u})\cdot\boldsymbol{v} - (\boldsymbol{\nabla}\times\boldsymbol{v})\cdot\boldsymbol{u}$.
- $\boldsymbol{\nabla}\times(\boldsymbol{u}\times\boldsymbol{v}) = \boldsymbol{u}(\boldsymbol{\nabla}\cdot\boldsymbol{v}) + \boldsymbol{v}\cdot\boldsymbol{\nabla}\boldsymbol{u} - \boldsymbol{u}\cdot\boldsymbol{\nabla}\boldsymbol{v} - \boldsymbol{v}(\boldsymbol{\nabla}\cdot\boldsymbol{u})$.
- $\boldsymbol{\nabla}(\boldsymbol{u}\cdot\boldsymbol{v}) = \boldsymbol{u}\times(\boldsymbol{\nabla}\times\boldsymbol{v}) + \boldsymbol{v}\times(\boldsymbol{\nabla}\times\boldsymbol{u}) + \boldsymbol{u}\cdot\boldsymbol{\nabla}\boldsymbol{v} + \boldsymbol{v}\cdot\boldsymbol{\nabla}\boldsymbol{u}$.

EXERCISES

4.9 Write in suffix notation the vector equation $\boldsymbol{a}\times\boldsymbol{b}+\boldsymbol{c} = (\boldsymbol{a}\cdot\boldsymbol{b})\boldsymbol{b} - \boldsymbol{d}$.

4.10 Simplify the suffix notation expressions

(a) $\delta_{ij}\delta_{jk}\delta_{ki}$;

(b) $\epsilon_{ijk}\epsilon_{klm}\epsilon_{mni}$.

4.11 Simplify the suffix notation expression $\delta_{ij}a_jb_lc_k\delta_{li}$ and write the result in vector form.

4.12 (a) Show that $\boldsymbol{\nabla}\times(f\boldsymbol{\nabla} f) = \boldsymbol{0}$.

(b) Evaluate $\boldsymbol{\nabla}\cdot(f\boldsymbol{\nabla} f)$.

4.13 Show that the vector $\boldsymbol{u} = \boldsymbol{\nabla} f\times\boldsymbol{\nabla} g$ is solenoidal.

4.14 Verify the formula (4.34) for $\boldsymbol{u}\cdot\boldsymbol{\nabla}\boldsymbol{u}$ by using (4.35).

4.15 Show that $\boldsymbol{\nabla}\cdot\nabla^2\boldsymbol{u} = \nabla^2\boldsymbol{\nabla}\cdot\boldsymbol{u}$,

(a) using suffix notation;

(b) using (4.24).

4.16 The vector fields $\boldsymbol{u}$ and $\boldsymbol{w}$ and the scalar field ϕ are related by the equation

$$\boldsymbol{u} + \boldsymbol{\nabla}\times\boldsymbol{w} = \boldsymbol{\nabla}\phi + \nabla^2\boldsymbol{u},$$

and $\boldsymbol{u}$ is solenoidal. Show that ϕ obeys Laplace's equation.

4.17 Show that $\boldsymbol{\nabla} f(r) = f'(r)\boldsymbol{r}/r$, where $\boldsymbol{r}$ is the position vector $\boldsymbol{r} = (x_1, x_2, x_3)$ and $r = |\boldsymbol{r}|$.

4.18 The vector field $\boldsymbol{u}$ is defined by $\boldsymbol{u} = h(r)\boldsymbol{r}$, where $h(r)$ is an arbitrary differentiable function.

(a) Show that $\boldsymbol{\nabla}\times\boldsymbol{u} = \boldsymbol{0}$.

(b) If $\boldsymbol{\nabla}\cdot\boldsymbol{u} = 0$, find the differential equation satisfied by h.

(c) Solve this differential equation.

4.19 A vector field $\boldsymbol{u}$ with the property that $\boldsymbol{u} = c\boldsymbol{\nabla}\times\boldsymbol{u}$, where c is a constant, is called a *Beltrami field.*

(a) Show that a Beltrami field is solenoidal.

(b) Show that the curl of a Beltrami field is a Beltrami field.

(c) A Beltrami field has the form $\boldsymbol{u} = (\sin y, f, g)$. Find the functions f and g and the possible values of c if it is given that g does not depend on x.

5
Integral Theorems

This chapter describes two important theorems that link the material in Chapter 2 on line, surface and volume integrals with the definitions of the divergence and curl from Chapter 3. These theorems have great physical significance and are widely used in deriving mathematical equations representing physical laws.

5.1 Divergence theorem

Let $\boldsymbol{u}$ be a continuously differentiable vector field, defined in a volume V. Let S be the closed surface forming the boundary of V and let $\boldsymbol{n}$ be the unit outward normal to S. Then the divergence theorem states that

$$\iiint_V \boldsymbol{\nabla} \cdot \boldsymbol{u} \, dV = \oiint_S \boldsymbol{u} \cdot \boldsymbol{n} \, dS. \tag{5.1}$$

Proof

The volume V is divided into a large number of small subvolumes δV_i with surfaces δS_i, as shown in Figure 5.1. The proof of the divergence theorem then follows naturally from the physical definition of the divergence in terms of a surface integral (3.14). Within each of the subvolumes, $\boldsymbol{\nabla} \cdot \boldsymbol{u}$ is defined by

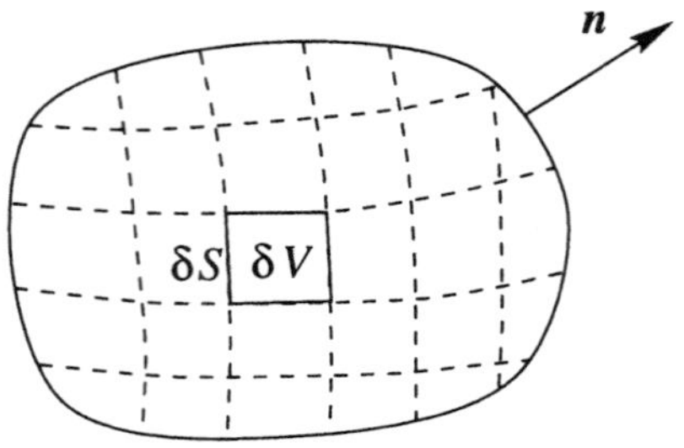

Fig. 5.1. Division of a volume V into small subvolumes δV for the proof of the divergence theorem.

$$\nabla \cdot \boldsymbol{u} \approx \frac{1}{\delta V_i} \oiint_{\delta S_i} \boldsymbol{u} \cdot \boldsymbol{n} \, dS, \tag{5.2}$$

where the approximation becomes exact in the limit $\delta V_i \to 0$. Now multiply both sides of (5.2) by δV_i and add the contributions from all the subvolumes:

$$\sum_i \nabla \cdot \boldsymbol{u} \, \delta V_i \approx \sum_i \oiint_{\delta S_i} \boldsymbol{u} \cdot \boldsymbol{n} \, dS. \tag{5.3}$$

Now take the limit $\delta V_i \to 0$. The l.h.s. becomes the volume integral of $\nabla \cdot \boldsymbol{u}$ over the volume V; this is just the definition of the volume integral. To simplify the r.h.s. consider two adjacent volume elements δV_1 and δV_2 (Figure 5.2). Since

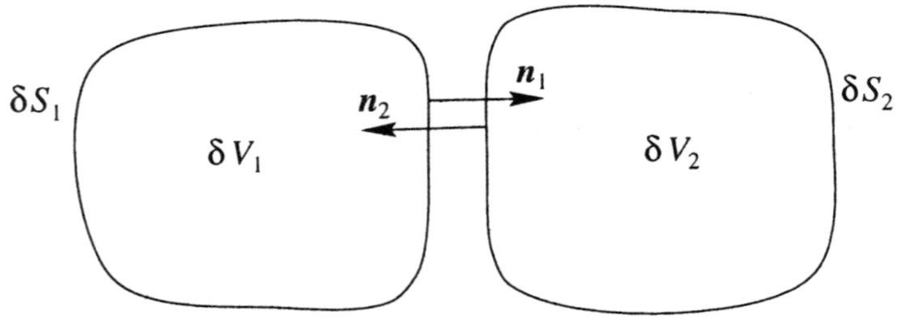

Fig. 5.2. Enlargement of two adjacent volume elements.

the normal vector to each surface points outward, the normal vectors to the two surfaces along their common surface point in opposite directions: $\boldsymbol{n}_1 = -\boldsymbol{n}_2$. Therefore the values of $\boldsymbol{u} \cdot \boldsymbol{n}$ cancel along the common surface: $\boldsymbol{u} \cdot \boldsymbol{n}_1 + \boldsymbol{u} \cdot \boldsymbol{n}_2 = 0$. This means that all the contributions to the sum on the r.h.s. of (5.3) from the interior of the region V cancel out, leaving only the surface integral over the exterior surface S. So in the limit $\delta V_i \to 0$, (5.3) becomes

$$\iiint_V \nabla \cdot \boldsymbol{u}\, dV = \oiint_S \boldsymbol{u} \cdot \boldsymbol{n}\, dS.$$

□

The divergence theorem is sometimes referred to as Gauss's theorem. It has many important applications in physics, and it is important to develop a physical intuition for the meaning of the theorem. Roughly speaking, the divergence theorem states that the total amount of expansion of $\boldsymbol{u}$ within the volume V is equal to the flux of $\boldsymbol{u}$ out of the surface S. This is essentially a conservation law, and the mathematical form of many physical conservation laws is derived from the divergence theorem. An example is given in the following section.

5.1.1 Conservation of mass for a fluid

As an example of the application of the divergence theorem, this section presents the derivation of the law of conservation of mass for a fluid of variable density.

Consider a fluid with density $\rho(\boldsymbol{r}, t)$ flowing with velocity $\boldsymbol{u}(\boldsymbol{r}, t)$. Let V be an arbitrary volume fixed in space, with surface S and outward normal $\boldsymbol{n}$ (Figure 5.3). Then the total mass of the fluid contained in V is the volume integral of ρ:

$$\text{Mass of fluid in } V = \iiint_V \rho\, dv. \tag{5.4}$$

Now the rate at which mass enters V is equal to the surface integral of the flux $\rho\,\boldsymbol{u}$:

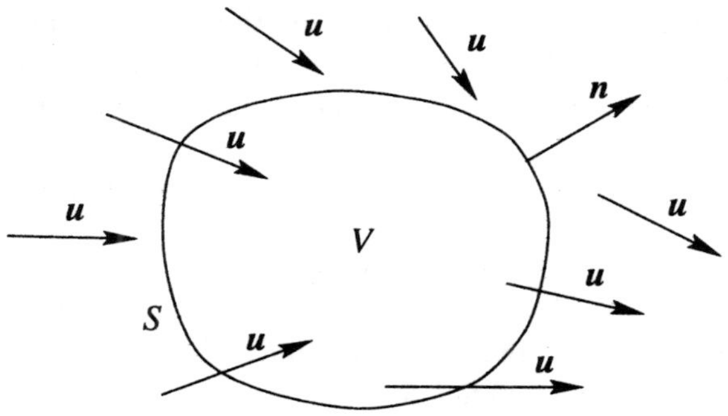

Fig. 5.3. Fluid flows with velocity $\boldsymbol{u}$ through a region V.

$$\text{Rate of mass flow into } V = -\oiint_S \rho\, \boldsymbol{u} \cdot \boldsymbol{n}\, dS, \tag{5.5}$$

where the minus sign appears because $\boldsymbol{n}$ points outward, so mass enters V if $\boldsymbol{u} \cdot \boldsymbol{n} < 0$.

We can now apply the physical law that mass is conserved: the rate of change of the mass in V must equal the rate at which mass enters V. Mathematically this is

$$\frac{d}{dt} \iiint_V \rho\, dV = -\oiint_S \rho\, \boldsymbol{u} \cdot \boldsymbol{n}\, dS. \tag{5.6}$$

The surface integral on the r.h.s. can now be written as a volume integral using the divergence theorem. Also, the order of the derivative and the integral on the l.h.s. can be interchanged:

$$\iiint_V \frac{\partial \rho}{\partial t}\, dV = -\iiint_V \boldsymbol{\nabla} \cdot (\rho \boldsymbol{u})\, dV, \tag{5.7}$$

where the time derivative has become a partial derivative since ρ is a function of space and time. These two integrals can now be combined into one:

$$\iiint_V \frac{\partial \rho}{\partial t} + \boldsymbol{\nabla} \cdot (\rho \boldsymbol{u})\ dV = 0. \tag{5.8}$$

Now this result has been obtained without any restrictions on the volume V. Thus it is true for any arbitrary volume V. The only way that this can be true is if the integrand (the quantity inside the integral) is zero everywhere. If there were some point where the integrand were non-zero, a small volume could be drawn around that point, which would contradict (5.8).

Therefore the law for conservation of mass of a fluid is

$$\frac{\partial \rho}{\partial t} + \boldsymbol{\nabla} \cdot (\rho \boldsymbol{u}) = 0. \tag{5.9}$$

This conservation law takes the following form: the rate of change of the density plus the divergence of the flux is zero. Many other conservation laws can also be written in this form, for example conservation of energy or conservation of electric charge.

By expanding the divergence of $\rho \boldsymbol{u}$, (5.9) can be written in the form

$$\frac{\partial \rho}{\partial t} + \boldsymbol{u} \cdot \boldsymbol{\nabla} \rho + \rho \boldsymbol{\nabla} \cdot \boldsymbol{u} = 0. \tag{5.10}$$

If the density of the fluid is constant and uniform, i.e. independent of time and space, then this equation simplifies to

$$\boldsymbol{\nabla} \cdot \boldsymbol{u} = 0. \tag{5.11}$$

A fluid obeying (5.11) is said to be *incompressible*.

5.1.2 Applications of the divergence theorem

The divergence theorem has many important applications, in addition to the derivation of the mathematical form of conservation laws shown in the previous section. It can be used to simplify the evaluation of integrals, by converting a complicated volume integral into a simpler surface integral or vice versa. It can also be used to prove some important results, such as the uniqueness of the solution to Laplace's equation

$$\nabla^2 \phi = 0. \tag{5.12}$$

Some of these applications are illustrated in the following examples.

Example 5.1

Show that for any closed surface S,

$$\oiint_S (\boldsymbol{\nabla} \times \boldsymbol{u}) \cdot \boldsymbol{n} \, dS = 0.$$

Using the divergence theorem, the surface integral can be converted into a volume integral:

$$\oiint_S (\boldsymbol{\nabla} \times \boldsymbol{u}) \cdot \boldsymbol{n} \, dS = \iiint_V \boldsymbol{\nabla} \cdot (\boldsymbol{\nabla} \times \boldsymbol{u}) \, dV.$$

Since the combination div curl is always zero, this integral is zero.

Example 5.2

Find the relationship between the surface integral

$$\oiint_S \boldsymbol{r} \cdot \boldsymbol{n} \, dS$$

and the volume V contained within the closed surface S.

Applying the divergence theorem,

$$\oiint_S \boldsymbol{r} \cdot \boldsymbol{n} \, dS = \iiint_V \boldsymbol{\nabla} \cdot \boldsymbol{r} \, dV = \iiint_V 3 \, dV = 3V,$$

using the result (3.17) that $\boldsymbol{\nabla} \cdot \boldsymbol{r} = 3$. Thus the surface integral is three times the volume V.

Example 5.3

The scalar field ϕ obeys Laplace's equation (5.12) in the region V and obeys $\phi = 0$ on the surface S that encloses V. Show that the only possible solution for ϕ is $\phi = 0$ everywhere within V.

$$\nabla^2\phi = 0 \quad \Rightarrow \quad \phi\nabla^2\phi = 0 \quad \Rightarrow \quad \boldsymbol{\nabla}\cdot(\phi\boldsymbol{\nabla}\phi) - \boldsymbol{\nabla}\phi\cdot\boldsymbol{\nabla}\phi = 0,$$

using (4.27). Now integrate over the volume V and use the divergence theorem to convert the first term to a surface integral:

$$\oiint_S \phi\boldsymbol{\nabla}\phi\cdot\boldsymbol{n}\,dS - \iiint_V |\boldsymbol{\nabla}\phi|^2\,dV = 0.$$

Since $\phi = 0$ on S, the surface integral vanishes. The quantity $|\boldsymbol{\nabla}\phi|^2$ is always greater than or equal to zero, so its integral can only be zero if $\boldsymbol{\nabla}\phi = \mathbf{0}$ everywhere. This means that ϕ must be a constant, and since $\phi = 0$ on S, this constant must be zero, so $\phi = 0$ everywhere within V.

Example 5.4

The scalar field ϕ obeys Laplace's equation (5.12) in the region V and the value of ϕ is given on the surface S that encloses V. Show that the solution to Laplace's equation is unique.

To prove uniqueness, suppose that there are two different solutions, ϕ_1 and ϕ_2, obeying $\nabla^2\phi_1 = 0$ and $\nabla^2\phi_2 = 0$ in V. Since the value of ϕ is specified on S, $\phi_1 = \phi_2$ on S. Now consider the function $\psi = \phi_1 - \phi_2$. This function also obeys Laplace's equation, since $\nabla^2(\phi_1 - \phi_2) = \nabla^2\phi_1 - \nabla^2\phi_2 = 0$. Moreover, $\psi = 0$ on S since $\phi_1 = \phi_2$ on S. Now we can apply the result of Example 5.3 to ψ: the only solution to $\nabla^2\psi = 0$ in V, $\psi = 0$ on S is $\psi = 0$ everywhere. Therefore $\phi_1 = \phi_2$ everywhere, so the solution is unique.

5.1.3 Related theorems linking surface and volume integrals

There are several other relationships between surface and volume integrals that can be derived from the divergence theorem by making different choices for the vector $\boldsymbol{u}$.

- Choose $\boldsymbol{u} = \boldsymbol{a}f$, where $\boldsymbol{a}$ is a constant vector and f is a scalar field. Then $\boldsymbol{\nabla}\cdot\boldsymbol{u} = f\,\boldsymbol{\nabla}\cdot\boldsymbol{a} + \boldsymbol{a}\cdot\boldsymbol{\nabla}f = \boldsymbol{a}\cdot\boldsymbol{\nabla}f$ since $\boldsymbol{a}$ is constant. Applying the divergence theorem gives

  $$\iiint_V \boldsymbol{a}\cdot\boldsymbol{\nabla}f\,dV = \oiint_S \boldsymbol{a}f\cdot\boldsymbol{n}\,dS.$$

  Since $\boldsymbol{a}$ is constant, it can be taken out of the integrals:

  $$\boldsymbol{a}\cdot\left(\iiint_V \boldsymbol{\nabla}f\,dV - \oiint_S f\boldsymbol{n}\,dS\right) = 0.$$

  Now since $\boldsymbol{a}$ is an arbitrary constant vector, this holds for any $\boldsymbol{a}$. This can only be true if the vector quantity within the large brackets is zero (for example, choosing $\boldsymbol{a} = \boldsymbol{e}_1, \boldsymbol{e}_2, \boldsymbol{e}_3$ in turn shows that each of the components of the vector in large brackets is zero). The resulting integral theorem is then

$$\iiint_V \boldsymbol{\nabla} f \, dV = \oiint_S f\boldsymbol{n} \, dS. \tag{5.13}$$

- Choose $\boldsymbol{u} = \boldsymbol{a} \times \boldsymbol{v}$, where $\boldsymbol{a}$ is a constant vector and $\boldsymbol{v}$ is a vector field. Then $\boldsymbol{\nabla} \cdot \boldsymbol{u} = (\boldsymbol{\nabla} \times \boldsymbol{a}) \cdot \boldsymbol{v} - (\boldsymbol{\nabla} \times \boldsymbol{v}) \cdot \boldsymbol{a} = -(\boldsymbol{\nabla} \times \boldsymbol{v}) \cdot \boldsymbol{a}$. The divergence theorem gives

$$\iiint_V -(\boldsymbol{\nabla} \times \boldsymbol{v}) \cdot \boldsymbol{a} \, dV = \oiint_S \boldsymbol{a} \times \boldsymbol{v} \cdot \boldsymbol{n} \, dS = \oiint_S \boldsymbol{a} \cdot \boldsymbol{v} \times \boldsymbol{n} \, dS,$$

using the rule that the dot and cross may be interchanged in a scalar triple product. As in the previous example, the dot product with $\boldsymbol{a}$ can be taken out of the integral sign and then cancelled, giving

$$\iiint_V -\boldsymbol{\nabla} \times \boldsymbol{v} \, dV = \oiint_S \boldsymbol{v} \times \boldsymbol{n} \, dS. \tag{5.14}$$

- Choose $\boldsymbol{u} = f\boldsymbol{\nabla} g$, where f and g are two scalar fields. Then $\boldsymbol{\nabla} \cdot \boldsymbol{u} = \boldsymbol{\nabla} f \cdot \boldsymbol{\nabla} g + f\nabla^2 g$, and the divergence theorem gives

$$\iiint_V \boldsymbol{\nabla} f \cdot \boldsymbol{\nabla} g + f\nabla^2 g \, dV = \oiint_S f\boldsymbol{\nabla} g \cdot \boldsymbol{n} \, dS. \tag{5.15}$$

This result is known as Green's First Identity.

- Choose $\boldsymbol{u} = f\boldsymbol{\nabla} g - g\boldsymbol{\nabla} f$. By interchanging f and g in (5.15) and subtracting, we obtain

$$\iiint_V f\nabla^2 g - g\nabla^2 f \, dV = \oiint_S (f\boldsymbol{\nabla} g - g\boldsymbol{\nabla} f) \cdot \boldsymbol{n} \, dS, \tag{5.16}$$

which is known as Green's Second Identity.

Historical note

George Green (1793–1841) was a Nottingham miller who spent less than two years at school and learnt his mathematics by studying library books. In 1828 he published privately his first and greatest work, 'An essay on the application of mathematical analysis to the theories of electricity and magnetism', which includes the two theorems above. As with many geniuses his work was not appreciated until several years after his death. Green's mill in Nottingham has now been restored and is open to the public along with a Science Centre illustrating some of the applications of his work.

EXERCISES

5.1 Use the divergence theorem to evaluate the surface integral

$$\oiint_S \boldsymbol{u} \cdot \boldsymbol{n} \, dS$$

where $\boldsymbol{u} = (x \sin y, \cos^2 x, y^2 - z \sin y)$ and S is the surface of the sphere $x^2 + y^2 + (z-2)^2 = 1$.

5.2 Verify the divergence theorem, by calculating both the volume integral and the surface integral, for the vector field $\boldsymbol{u} = (y, x, z - x)$ and the volume V given by the unit cube $0 \le x, y, z, \le 1$.

5.3 An incompressible fluid is contained within a volume V with surface S and $\boldsymbol{u} \cdot \boldsymbol{n} = 0$ on S. Using the divergence theorem, show that

$$\iiint_V \boldsymbol{u} \cdot \boldsymbol{\nabla}\phi \, dV = 0$$

for any differentiable scalar field ϕ.

5.4 Two scalar fields f and g are related by Poisson's equation, $\nabla^2 f = g$. Show that

$$\iiint g \, dV = \oiint \boldsymbol{\nabla} f \cdot \boldsymbol{n} \, dS.$$

5.5 Use the divergence theorem to evaluate the surface integral

$$\iint_S \boldsymbol{v} \cdot \boldsymbol{n} \, dS$$

where $\boldsymbol{v} = (x + y, z^2, x^2)$ and S is the surface of the hemisphere $x^2 + y^2 + z^2 = 1$ with $z > 0$ and $\boldsymbol{n}$ is the upward-pointing normal. Note that the surface S is not closed.

5.6 Following the argument of Section 5.1.1, obtain the equation for conservation of electric charge relating the charge density q and the electric current density $\boldsymbol{j}$.

5.7 Use (5.13) to obtain a definition for $\boldsymbol{\nabla} f$ as the limit of an integral, similar to the definitions of div and curl.

5.2 Stokes's theorem

Stokes's theorem gives an alternative expression for the surface integral of the curl of a vector field. This is analogous to the divergence theorem, so Stokes's theorem could be referred to as the 'curl theorem'. The proof of the theorem is very similar to that for the divergence theorem, being based on the definition of curl in terms of a line integral.

Let C be a closed curve which forms the boundary of a surface S. Then for a continuously differentiable vector field $\boldsymbol{u}$, Stokes's theorem states that

$$\iint_S \nabla \times \boldsymbol{u} \cdot \boldsymbol{n}\, dS = \oint_C \boldsymbol{u} \cdot d\boldsymbol{r}, \tag{5.17}$$

where the direction of the line integral around C and the normal $\boldsymbol{n}$ are oriented in a right-handed sense (Figure 5.4).

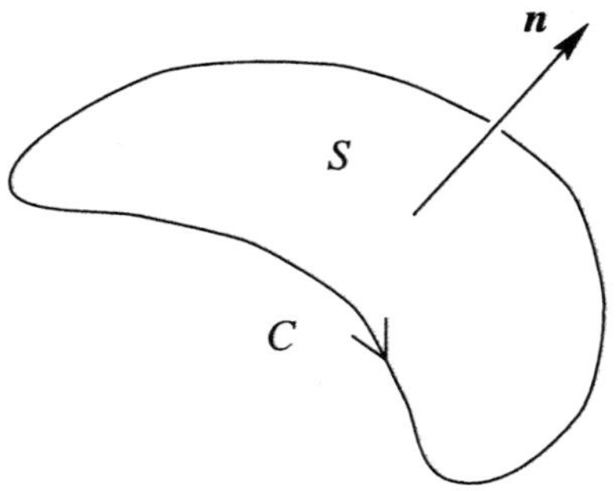

Fig. 5.4. Orientation of the curve C and the surface S for Stokes's theorem.

Proof

To demonstrate the theorem we first divide the surface S into small pieces each with area δS_i and bounding curves δC_i (Figure 5.5). Within each piece of the surface, the definition (3.20) of $\nabla \times \boldsymbol{u}$ is

$$\nabla \times \boldsymbol{u} \cdot \boldsymbol{n} \approx \frac{1}{\delta S_i} \oint_{\delta C_i} \boldsymbol{u} \cdot d\boldsymbol{r},$$

where the approximation is exact in the limit $\delta S_i \to 0$. Multiplying by δS_i and adding the contributions from all the surface elements,

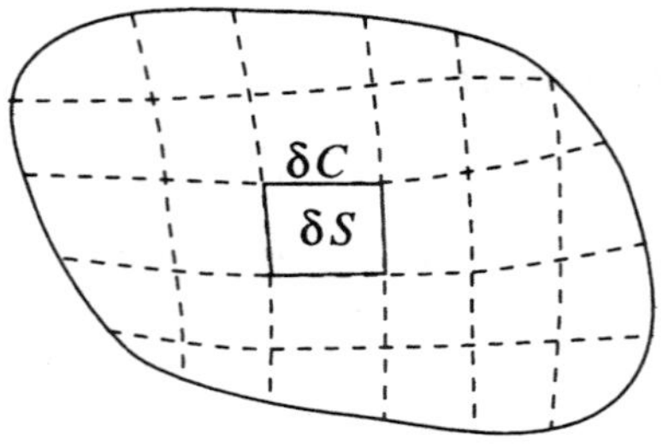

Fig. 5.5. Division of the surface S into small elements δS for the proof of Stokes's theorem.

$$\sum_i \boldsymbol{\nabla} \times \boldsymbol{u} \cdot \boldsymbol{n}\, \delta S_i \approx \sum_i \oint_{\delta C_i} \boldsymbol{u} \cdot \boldsymbol{dr}.$$

Now consider the limit $\delta S_i \to 0$. The l.h.s. gives the surface integral of $\boldsymbol{\nabla} \times \boldsymbol{u} \cdot \boldsymbol{n}$ over the surface S. On the r.h.s. the contributions to the line integrals from neighbouring elements cancel out, because the line elements $\boldsymbol{dr}$ point in opposite directions (Figure 5.6). Therefore only the curves that form part of C

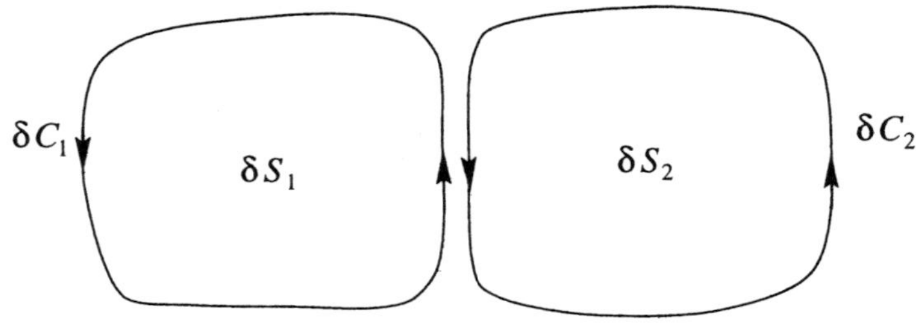

Fig. 5.6. Line integrals along adjoining elements cancel out.

contribute to the sum, so the sum simplifies to the line integral around C:

$$\iint_S \boldsymbol{\nabla} \times \boldsymbol{u} \cdot \boldsymbol{n}\, dS = \oint_C \boldsymbol{u} \cdot \boldsymbol{dr}.$$

□

5.2.1 Applications of Stokes's theorem

Stokes's theorem can be useful for evaluating integrals, by converting line integrals to surface integrals or vice versa. It can also be used to prove other theorems, as in Example 5.5 below, or to formulate physical laws (Example 5.6).

Example 5.5

Show that any irrotational vector field is conservative.

Suppose that $\boldsymbol{u}$ is irrotational, so $\boldsymbol{\nabla} \times \boldsymbol{u} = \boldsymbol{0}$. Then for any closed curve C,

$$\oint_C \boldsymbol{u} \cdot d\boldsymbol{r} = \iint_S \boldsymbol{\nabla} \times \boldsymbol{u} \cdot \boldsymbol{n}\, dS = 0,$$

where S is any surface spanning C. Thus $\boldsymbol{u}$ is a conservative vector field. Note that this result completes the demonstration of the statement in Section 3.4.3 of the equivalence of the three properties (i) $\boldsymbol{u} = \boldsymbol{\nabla}\phi$, (ii) $\boldsymbol{\nabla} \times \boldsymbol{u} = \boldsymbol{0}$, (iii) $\boldsymbol{u}$ is conservative.

Example 5.6

Ampère's law states that the total flux of electric current flowing through a loop is proportional to the line integral of the magnetic field around the loop. Use Stokes's theorem to obtain an alternative form of this law that does not involve any integrals.

Let $\boldsymbol{B}$ be the magnetic field strength and $\boldsymbol{j}$ be the current density. The constant of proportionality is μ_0 in SI units. Then Ampère's law states that

$$\oint_C \boldsymbol{B} \cdot d\boldsymbol{r} = \mu_0 \iint_S \boldsymbol{j} \cdot \boldsymbol{n}\, dS$$

for any surface S that spans the loop C. Using Stokes's theorem to transform the l.h.s. gives

$$\iint_S \boldsymbol{\nabla} \times \boldsymbol{B} \cdot \boldsymbol{n}\, dS = \mu_0 \iint_S \boldsymbol{j} \cdot \boldsymbol{n}\, dS.$$

Now if this is true for any loop C, and so any surface S, it follows that

$$\boldsymbol{\nabla} \times \boldsymbol{B} = \mu_0 \boldsymbol{j}.$$

Note the similarity between this argument and that used when applying the divergence theorem to the conservation of mass of a fluid in Section 5.1.1.

Example 5.7

Use Stokes's theorem to show that for any closed surface S,

$$\oiint_S (\nabla \times \boldsymbol{u}) \cdot \boldsymbol{n} \, dS = 0.$$

Consider the case where a small hole is made in the closed surface. Then by Stokes's theorem, the surface integral of $(\nabla \times \boldsymbol{u}) \cdot \boldsymbol{n}$ over the surface S is equal to the line integral of $\boldsymbol{u} \cdot \boldsymbol{dr}$ around the perimeter of the small hole. As the size of the hole shrinks to zero, so does the value of the line integral, giving the required result. Note that this result was obtained using the divergence theorem in Example 5.1.

Example 5.8

The surface S is defined by $x^2 + 4y^2 = 1$, $-1 \le z \le 1$. Use Stokes's theorem to evaluate the surface integral

$$\iint_S (xz^2, -yz^2, 0) \cdot \boldsymbol{n} \, dS.$$

Note that this surface is not simply connected, but Stokes's theorem can still be applied. By imagining a cut in the surface (Figure 5.7), the surface integral is equal to the sum of two line integrals around the two elliptical curves C_1 and C_2 that form the ends of the cylindrical surface. In order to apply Stokes's theorem, the vector field $(xz^2, -yz^2, 0)$ must be written as the curl of another vector field $\boldsymbol{u}$. Seeking a solution of the form $\boldsymbol{u} = (0, 0, h(x, y, z))$, this can be

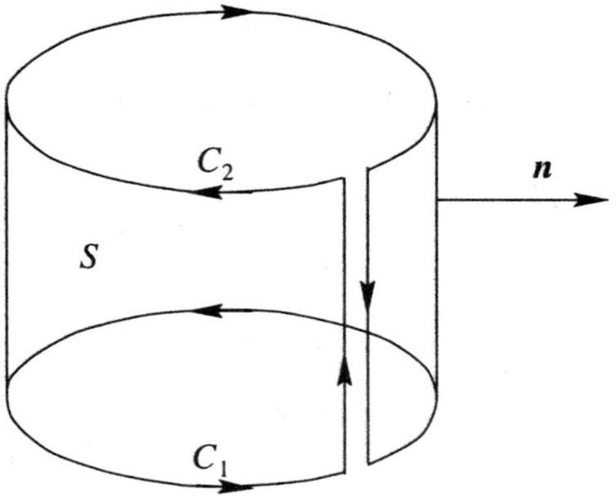

Fig. 5.7. Stokes's theorem can be used to transform the surface integral over the curved surface of the cylinder into two line integrals around the ends of the cylinder, by introducing a cut in the surface. The two line integrals along the cut cancel out.

achieved if

$$xz^2 = \frac{\partial h}{\partial y}, \qquad -yz^2 = -\frac{\partial h}{\partial x}$$

and these two equations are satisfied if $h = xyz^2$. Now since the two curves C_1 and C_2 lie in planes $z =$ constant, so $\boldsymbol{dr} = (dx, dy, 0)$, $\boldsymbol{u} \cdot \boldsymbol{dr} = 0$ on C_1 and C_2 so the value of the integral is zero.

5.2.2 Related theorems linking line and surface integrals

As in the case of the divergence theorem, Stokes's theorem can be used to derive other theorems that relate line integrals to surface integrals by appropriate choices of the vector field $\boldsymbol{u}$.

- Choose $\boldsymbol{u} = \boldsymbol{a}f$, where $\boldsymbol{a}$ is a constant vector and f is a scalar field. Then $\boldsymbol{\nabla} \times \boldsymbol{u} = \boldsymbol{\nabla} f \times \boldsymbol{a} + f\boldsymbol{\nabla} \times \boldsymbol{a} = \boldsymbol{\nabla} f \times \boldsymbol{a}$, so Stokes's theorem gives

$$\iint_S \boldsymbol{\nabla} f \times \boldsymbol{a} \cdot \boldsymbol{n}\; dS = \oint_C \boldsymbol{a} f \cdot \boldsymbol{dr}.$$

  Using the rules for manipulating the scalar triple product and taking out the constant vector $\boldsymbol{a}$ from the integrals gives

$$\boldsymbol{a} \cdot \left(\iint_S -\boldsymbol{\nabla} f \times \boldsymbol{n}\; dS \right) = \boldsymbol{a} \cdot \left(\oint_C f\, \boldsymbol{dr} \right).$$

  As in Section 5.1.3, the constant $\boldsymbol{a}$ can be cancelled, giving

$$\iint_S -\boldsymbol{\nabla} f \times \boldsymbol{n}\; dS = \oint_C f\, \boldsymbol{dr}.$$

- Choose $\boldsymbol{u} = \boldsymbol{a} \times \boldsymbol{v}$, where $\boldsymbol{a}$ is a constant vector and $\boldsymbol{v}$ is a vector field. This case is more complicated, but provides a good example for the use of suffix notation. The line integral in Stokes's theorem is

$$\oint_C \boldsymbol{a} \times \boldsymbol{v} \cdot \boldsymbol{dr} = \boldsymbol{a} \cdot \oint_C \boldsymbol{v} \times \boldsymbol{dr},$$

  interchanging the dot and the cross and taking the constant $\boldsymbol{a}$ outside the integral. Since $\boldsymbol{\nabla} \times (\boldsymbol{a} \times \boldsymbol{v}) = \boldsymbol{a}(\boldsymbol{\nabla} \cdot \boldsymbol{v}) - \boldsymbol{a} \cdot \boldsymbol{\nabla} \boldsymbol{v}$, from formula (4.30), the surface integral is

$$\begin{aligned}
\iint_S (\boldsymbol{a}(\boldsymbol{\nabla} \cdot \boldsymbol{v}) - \boldsymbol{a} \cdot \boldsymbol{\nabla} \boldsymbol{v}) \cdot \boldsymbol{n}\; dS &= \iint_S \left(a_j \frac{\partial v_k}{\partial x_k} - a_k \frac{\partial v_j}{\partial x_k} \right) n_j\; dS \\
&= \iint_S a_j \left(\frac{\partial v_k}{\partial x_k} n_j - \frac{\partial v_k}{\partial x_j} n_k \right) dS
\end{aligned}$$

where j and k have been interchanged in the second term. Applying Stokes's theorem and cancelling the a_j gives a relationship between a surface integral and a line integral in suffix notation:

$$\iint_S \frac{\partial v_k}{\partial x_k} n_j - \frac{\partial v_k}{\partial x_j} n_k \, dS = \left[\oint_C \boldsymbol{v} \times d\boldsymbol{r} \right]_j . \tag{5.18}$$

To obtain the form of this equation in vector notation, consider the quantity $[(\boldsymbol{n} \times \boldsymbol{\nabla}) \times \boldsymbol{v}]_j$. In suffix notation this is

$$\begin{aligned} \epsilon_{jkl}\epsilon_{kmn} n_m \frac{\partial}{\partial x_n} v_l &= (\delta_{lm}\delta_{jn} - \delta_{ln}\delta_{jm}) n_m \frac{\partial v_l}{\partial x_n} \\ &= n_l \frac{\partial v_l}{\partial x_j} - n_j \frac{\partial v_l}{\partial x_l}. \end{aligned}$$

This is now minus the quantity appearing in the surface integral (5.18), so the vector form of (5.18) is

$$\iint_S -(\boldsymbol{n} \times \boldsymbol{\nabla}) \times \boldsymbol{v} \, dS = \oint_C \boldsymbol{v} \times d\boldsymbol{r}. \tag{5.19}$$

Note that suffix notation is the only secure method for obtaining such results, short of writing out all the components of the vector quantities longhand. Attempts to expand using the rules for a vector triple product generally give incorrect results.

- Choose the surface S to be a flat surface lying in the x, y plane, so $\boldsymbol{n} = (0, 0, 1)$ and choose $\boldsymbol{u} = (F(x,y), G(x,y), 0)$. Then

$$\boldsymbol{\nabla} \times \boldsymbol{u} = \left(0, 0, \frac{\partial G}{\partial x} - \frac{\partial F}{\partial y}\right)$$

and $\boldsymbol{u} \cdot d\boldsymbol{r} = F\,dx + G\,dy$. Then Stokes's theorem gives

$$\iint_S \frac{\partial G}{\partial x} - \frac{\partial F}{\partial y} \, dx\,dy = \oint_C F\,dx + G\,dy, \tag{5.20}$$

a result known as Green's theorem.

Summary of Chapter 5

- The divergence theorem states that

$$\iiint_V \boldsymbol{\nabla} \cdot \boldsymbol{u}\, dV = \oiint_S \boldsymbol{u} \cdot \boldsymbol{n}\, dS,$$

  where S is the surface enclosing the volume V and $\boldsymbol{n}$ is the outward-pointing unit normal vector.
- Geometrically, the divergence theorem follows naturally from the physical definition of the divergence.
- The divergence theorem has many applications, including simplifying the evaluation of surface or volume integrals, deriving physical conservation laws and showing that Laplace's equation has a unique solution.
- Stokes's theorem states that

$$\iint_S \boldsymbol{\nabla} \times \boldsymbol{u} \cdot \boldsymbol{n}\, dS = \oint_C \boldsymbol{u} \cdot \boldsymbol{dr},$$

  where the curve C encloses the surface S and C and $\boldsymbol{n}$ are oriented in a right-handed sense.
- Stokes's theorem follows from the definition of the curl in terms of a line integral.
- A number of other related theorems linking volume, surface and line integrals can be derived from the divergence theorem and Stokes's theorem.

EXERCISES

5.8 Show that

$$\oint_C \boldsymbol{r} \cdot d\boldsymbol{r} = 0$$

for any closed curve C.

5.9 Verify Stokes's theorem by evaluating both the line and surface integrals for the vector field $\boldsymbol{u} = (2x - y, -y^2, -y^2 z)$ and the surface S given by the disk $z = 0$, $x^2 + y^2 \leq 1$.

5.10 Use Stokes's theorem to show that

$$\oint_C f \boldsymbol{\nabla} g \cdot d\boldsymbol{r} = - \oint_C g \boldsymbol{\nabla} f \cdot d\boldsymbol{r}$$

for any closed curve C and differentiable scalar fields f and g.

5.11 If $\boldsymbol{u}$ is irrotational, express the surface integral

$$\iint_S \boldsymbol{u} \times \boldsymbol{\nabla} f \cdot \boldsymbol{n} \, dS$$

as a line integral.

5.12 The magnetic field $\boldsymbol{B}$ in an electrically conducting fluid moving with velocity $\boldsymbol{u}$ obeys the magnetic induction equation

$$\frac{\partial \boldsymbol{B}}{\partial t} = \boldsymbol{\nabla} \times (\boldsymbol{u} \times \boldsymbol{B}).$$

Show that the total flux of magnetic field through a surface enclosed by a streamline of the flow (a closed curve which is everywhere parallel to $\boldsymbol{u}$) is independent of time.

5.13 Use (5.18) to show that the area A of a flat surface S enclosed by a curve C is

$$A = 1/2 \left| \oint_C \boldsymbol{r} \times d\boldsymbol{r} \right| .$$

6
Curvilinear Coordinates

6.1 Orthogonal curvilinear coordinates

So far in this book we have used rectangular Cartesian coordinates. In many physical problems, however, these are not the most convenient coordinates to use. Consider, for example, the problem of finding the electric field produced by a charged sphere. In this chapter the general theory of non-Cartesian coordinate systems is introduced. Formulae for grad, div and curl in these coordinate systems are developed and the two most important examples, cylindrical and spherical polar coordinates are described.

Suppose a transformation is carried out from a Cartesian coordinate system (x_1, x_2, x_3) to another coordinate system (u_1, u_2, u_3). This new system will be called a *curvilinear coordinate system.* It will be assumed that there is a one-to-one relationship between the x_i and the u_i, so that for example x_1 can be written as a function of the u_i, $x_1 = x_1(u_1, u_2, u_3)$ and conversely $u_1 = u_1(x_1, x_2, x_3)$.

The surfaces u_i = constant are referred to as *coordinate surfaces* and the intersection of these surfaces defines the *coordinate curves*, so for example the u_1 coordinate curve is the intersection of the surfaces u_2 = constant and u_3 = constant (Figure 6.1).

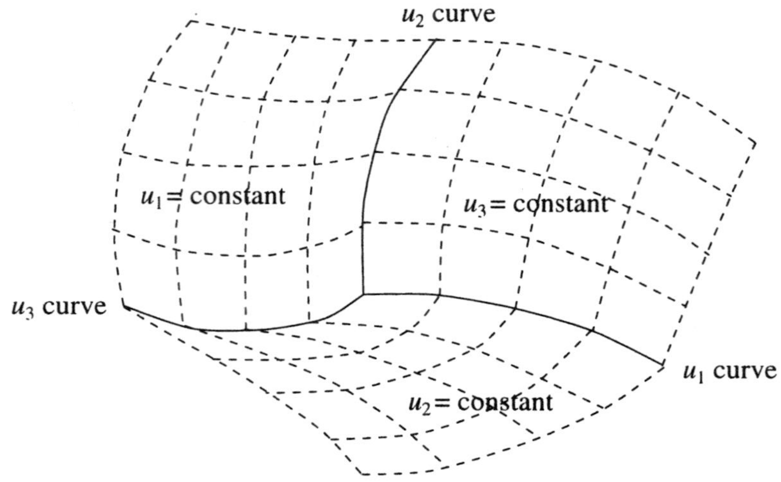

Fig. 6.1. The coordinate surfaces and coordinate curves of a curvilinear coordinate system.

Now consider a small displacement $\boldsymbol{dx} = (dx_1, dx_2, dx_3)$. Since the x_i are functions of the u_i this can be written as

$$\boldsymbol{dx} = \frac{\partial \boldsymbol{x}}{\partial u_1} du_1 + \frac{\partial \boldsymbol{x}}{\partial u_2} du_2 + \frac{\partial \boldsymbol{x}}{\partial u_3} du_3,$$

or more compactly using suffix notation as

$$dx_i = \frac{\partial x_i}{\partial u_j} du_j,$$

where the repeated suffix j on the r.h.s. implies summation from 1 to 3. Now the partial derivative $\partial \boldsymbol{x}/\partial u_1$ means the rate of variation of $\boldsymbol{x}$ with u_1 while u_2 and u_3 are held fixed, so the vector $\partial \boldsymbol{x}/\partial u_1$ lies in the u_2 and u_3 coordinate surfaces and is therefore tangent to the u_1 coordinate curve. This enables a unit vector $\boldsymbol{e_1}$ to be defined in the direction of the u_1 curve, by

$$\boldsymbol{e_1} = \frac{\partial \boldsymbol{x}}{\partial u_1} \Big/ h_1 \tag{6.1}$$

where h_1 is a *scale factor* defined by

$$h_1 = \left| \frac{\partial \boldsymbol{x}}{\partial u_1} \right| . \tag{6.2}$$

The unit vectors $\boldsymbol{e_2}$ and $\boldsymbol{e_3}$ are defined in a similar way, along with the scale factors h_2 and h_3.

The displacement vector $\boldsymbol{dx}$ can then be written in terms of these unit vectors and scale factors as

$$\boldsymbol{dx} = h_1 \boldsymbol{e_1} du_1 + h_2 \boldsymbol{e_2} du_2 + h_3 \boldsymbol{e_3} du_3.$$

Attention will be restricted to coordinate systems in which the unit vectors are orthogonal, so that

$$\boldsymbol{e_i} \cdot \boldsymbol{e_j} = \delta_{ij}. \tag{6.3}$$

Such coordinate systems are known as *orthogonal curvilinear coordinates.* This means that the coordinate curves are perpendicular to each other where they intersect. It will also be assumed that the coordinate system is right-handed, so that

$$\boldsymbol{e_1} \times \boldsymbol{e_2} = \boldsymbol{e_3}. \tag{6.4}$$

Locally, this coordinate system appears as a rectangular coordinate system with axes stretched by the factors h_i, so that a change in u_1 of size du_1 leads to a change of distance $h_1\, du_1$ in the $\boldsymbol{e_1}$ direction (Figure 6.2). Globally, however, the directions of the unit vectors $\boldsymbol{e_i}$ vary in space, as do the scale factors h_i.

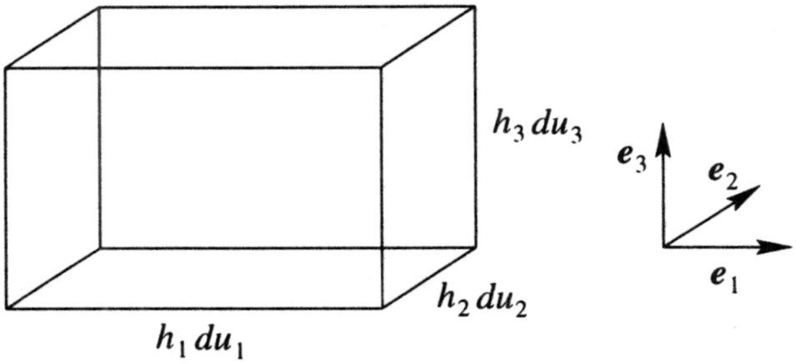

Fig. 6.2. An orthogonal curvilinear coordinate system appears locally as a rectangular coordinate system.

Having set up this general framework, formulae for various useful quantities can be derived in terms of the scale factors h_i with reference to Figure 6.2. The length of a line element ds is found from

$$ds^2 = \boldsymbol{dx} \cdot \boldsymbol{dx} = h_1^2 du_1^2 + h_2^2 du_2^2 + h_3^2 du_3^2. \tag{6.5}$$

A surface element dS on the u_1 coordinate surface generated by displacements du_2, du_3 is rectangular and so has the area

$$dS = h_2 h_3\, du_2\, du_3, \tag{6.6}$$

and similarly for area elements on the u_2 and u_3 coordinate surfaces. Finally, the volume element dV produced by displacements du_1, du_2, du_3 is again rectangular and so its volume is

$$dV = h_1 h_2 h_3 \, du_1 \, du_2 \, du_3. \tag{6.7}$$

Example 6.1

Show that the volume element in a right-handed orthogonal curvilinear coordinate system, $h_1 h_2 h_3 \, du_1 \, du_2 \, du_3$, can also be written in terms of the Jacobian of the transformation, J, defined to be the determinant of the matrix with i, j element $\partial x_i / \partial u_j$.

The determinant of a matrix can be interpreted as the scalar triple product of the vectors forming its rows or columns, so J can be written as the scalar triple product of three vectors:

$$\begin{aligned} J &= \frac{\partial \boldsymbol{x}}{\partial u_1} \cdot \frac{\partial \boldsymbol{x}}{\partial u_2} \times \frac{\partial \boldsymbol{x}}{\partial u_3} \\ &= h_1 \boldsymbol{e}_1 \cdot h_2 \boldsymbol{e}_2 \times h_3 \boldsymbol{e}_3 \\ &= h_1 h_2 h_3 \end{aligned}$$

since for a right-handed coordinate system, $\boldsymbol{e}_2 \times \boldsymbol{e}_3 = \boldsymbol{e}_1$ so $\boldsymbol{e}_1 \cdot \boldsymbol{e}_2 \times \boldsymbol{e}_3 = 1$. Therefore the volume element can be written

$$dV = J \, du_1 \, du_2 \, du_3.$$

Example 6.2

Parabolic coordinates (u, v, w) are related to Cartesian coordinates (x_1, x_2, x_3) by the equations

$$x_1 = 2uv, \quad x_2 = u^2 - v^2, \quad x_3 = w.$$

Sketch the u and v coordinate curves, find the scale factors h_u, h_v, h_w and the unit vectors $\boldsymbol{e_u}$, $\boldsymbol{e_v}$, $\boldsymbol{e_w}$, and check that the (u, v, w) coordinate system is orthogonal.

The u and v coordinate curves are the intersections of the v and u coordinate surfaces with the w coordinate surfaces. Since $w = x_3$, the x_1, x_2 plane is a w coordinate surface. Consider the surface $u = c$, where c is a constant. In terms of the Cartesian coordinates this can be written $x_2 = c^2 - x_1^2/4c^2$ after eliminating v. Similarly the surface $v = k$ can be written $x_2 = -k^2 + x_1^2/4k^2$. The u and v coordinate curves are therefore parabolas in the x_1, x_2 plane (Figure 6.3).

The scale factors h_u, h_v, h_w are just the magnitudes of the partial derivatives of the vector (x_1, x_2, x_3) with respect to u, v and w:

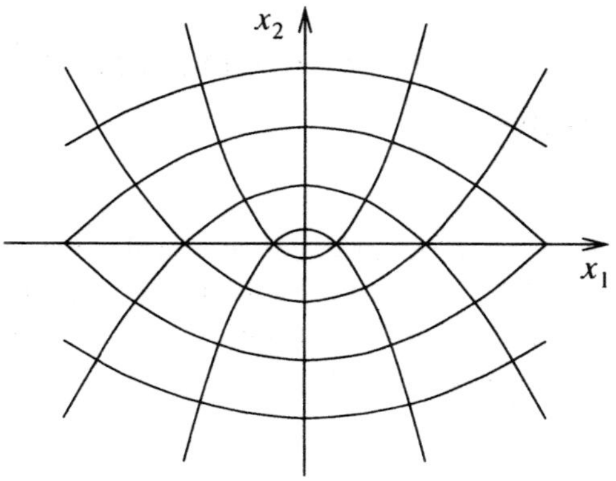

Fig. 6.3. The coordinate curves of the parabolic coordinate system.

$$\begin{aligned} h_u &= |(2v, 2u, 0)| = 2\sqrt{u^2 + v^2}, \\ h_v &= |(2u, -2v, 0)| = 2\sqrt{u^2 + v^2}, \\ h_w &= |(0, 0, 1)| = 1. \end{aligned}$$

The unit vectors are the vectors of the partial derivatives divided by the scale factors:

$$\begin{aligned} \boldsymbol{e_u} &= (v, u, 0)/\sqrt{u^2 + v^2}, \\ \boldsymbol{e_v} &= (u, -v, 0)/\sqrt{u^2 + v^2}, \\ \boldsymbol{e_w} &= (0, 0, 1). \end{aligned}$$

Since the dot product of any two of these unit vectors is zero, the (u, v, w) system is orthogonal.

6.2 Grad, div and curl in orthogonal curvilinear coordinate systems

In this section, formulae for the gradient of a scalar field and the divergence and curl of a vector field are derived for orthogonal curvilinear coordinate systems. In each case the physical definition is used, i.e. the definition which is independent of any coordinate system.

6.2.1 Gradient

The gradient ∇f of a scalar field f is a vector perpendicular to the surfaces $f =$ constant, defined by the equation

$$df = \nabla f \cdot \boldsymbol{dx}$$

where $\boldsymbol{dx}$ is an infinitesimal change in position and df is the corresponding change in f. Now from (6.1) we have

$$\boldsymbol{dx} = h_1 \boldsymbol{e_1} du_1 + h_2 \boldsymbol{e_2} du_2 + h_3 \boldsymbol{e_3} du_3.$$

If f is written as a function of the u_i then

$$\begin{aligned} df &= \frac{\partial f}{\partial u_1} du_1 + \frac{\partial f}{\partial u_2} du_2 + \frac{\partial f}{\partial u_3} du_3 \\ &= \frac{1}{h_1}\frac{\partial f}{\partial u_1} h_1 du_1 + \frac{1}{h_2}\frac{\partial f}{\partial u_2} h_2 du_2 + \frac{1}{h_3}\frac{\partial f}{\partial u_3} h_3 du_3 \\ &= \left(\frac{1}{h_1}\frac{\partial f}{\partial u_1} \boldsymbol{e_1} + \frac{1}{h_2}\frac{\partial f}{\partial u_2} \boldsymbol{e_2} + \frac{1}{h_3}\frac{\partial f}{\partial u_3} \boldsymbol{e_3} \right) \cdot \boldsymbol{dx}. \end{aligned}$$

Since this holds for any $\boldsymbol{dx}$, the term in the large brackets is ∇f:

$$\nabla f = \frac{1}{h_1}\frac{\partial f}{\partial u_1} \boldsymbol{e_1} + \frac{1}{h_2}\frac{\partial f}{\partial u_2} \boldsymbol{e_2} + \frac{1}{h_3}\frac{\partial f}{\partial u_3} \boldsymbol{e_3}. \qquad (6.8)$$

6.2.2 Divergence

The formula for the divergence of a vector field $\boldsymbol{v} = v_1\boldsymbol{e_1} + v_2\boldsymbol{e_2} + v_3\boldsymbol{e_3}$ in orthogonal curvilinear coordinates can be obtained using the definition (3.14)

$$\boldsymbol{\nabla} \cdot \boldsymbol{v} = \lim_{\delta V \to 0} \frac{1}{\delta V} \oiint_{\delta S} \boldsymbol{v} \cdot \boldsymbol{n}\, dS.$$

Since the coordinate system is orthogonal, the argument of Section 3.3 based on choosing δV to be a small rectangular box can be repeated, with reference to Figure 6.2. The only difference in the argument is that the lengths of the sides of the box are scaled by the scale factors h_i. On the u_1 surface on the right of the box, $\boldsymbol{n} = \boldsymbol{e_1}$ so $\boldsymbol{v} \cdot \boldsymbol{n} = v_1$ and

$$\boldsymbol{v} \cdot \boldsymbol{n}\, dS \approx v_1 h_2 h_3\, du_2 du_3$$

where v_1, h_2 and h_3 are evaluated at $(u_1 + du_1/2, u_2, u_3)$. Similarly, on the opposite surface,

$$\boldsymbol{v} \cdot \boldsymbol{n}\, dS \approx -v_1 h_2 h_3\, du_2 du_3$$

evaluated at $(u_1 - du_1/2, u_2, u_3)$. Adding these two contributions and dividing by the volume $h_1 h_2 h_3 du_1 du_2 du_3$ (exactly as in Section 3.3) gives the contribution to $\boldsymbol{\nabla} \cdot \boldsymbol{v}$ from these two surfaces as

$$\frac{1}{h_1 h_2 h_3} \frac{\partial}{\partial u_1}(v_1 h_2 h_3).$$

Note that h_2 and h_3 cannot be cancelled out because in general they are functions of u_1. The contributions to $\boldsymbol{\nabla} \cdot \boldsymbol{v}$ from the other surfaces follow from cyclic permutation, so

$$\boldsymbol{\nabla} \cdot \boldsymbol{v} = \frac{1}{h_1 h_2 h_3}\left(\frac{\partial}{\partial u_1}(v_1 h_2 h_3) + \frac{\partial}{\partial u_2}(v_2 h_3 h_1) + \frac{\partial}{\partial u_3}(v_3 h_1 h_2)\right). \tag{6.9}$$

By combining the definitions of div (6.9) and grad (6.8), a formula for the Laplacian of a scalar field, $\nabla^2 f = \boldsymbol{\nabla} \cdot (\boldsymbol{\nabla} f)$ can be obtained:

$$\begin{aligned} \nabla^2 f \;=\; & \frac{1}{h_1 h_2 h_3}\left[\frac{\partial}{\partial u_1}\left(\frac{h_2 h_3}{h_1}\frac{\partial f}{\partial u_1}\right)\right. \\ & \left. + \frac{\partial}{\partial u_2}\left(\frac{h_3 h_1}{h_2}\frac{\partial f}{\partial u_2}\right) + \frac{\partial}{\partial u_3}\left(\frac{h_1 h_2}{h_3}\frac{\partial f}{\partial u_3}\right)\right]. \end{aligned}$$

6.2.3 Curl

The curl of a vector field $\boldsymbol{v}$ in orthogonal curvilinear coordinates is found using the definition

$$\boldsymbol{n}\cdot\boldsymbol{\nabla}\times\boldsymbol{v} = \lim_{\delta S\to 0}\frac{1}{\delta S}\oint_{\delta C}\boldsymbol{v}\cdot d\boldsymbol{r}$$

and following the argument of Section 3.4. To find the $\boldsymbol{e_3}$ component of $\boldsymbol{\nabla}\times\boldsymbol{v}$, consider a small rectangle in the u_3 surface, with sides of length h_1du_1 and h_2du_2 (Figure 6.4). The line integral along the right-hand side of the rectangle

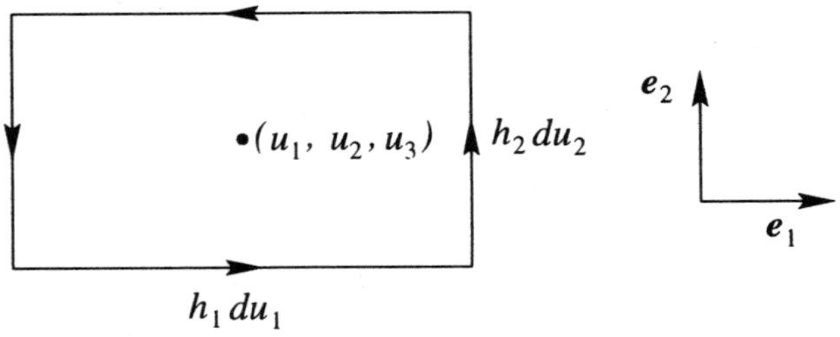

Fig. 6.4. Rectangle of four line segments for deriving the formula for $\boldsymbol{\nabla}\times\boldsymbol{v}$ in orthogonal curvilinear coordinates.

is approximately $v_2h_2du_2$ evaluated at $(u_1 + du_1/2, u_2, u_3)$ and the integral along the left side is approximately $-v_2h_2du_2$ evaluated at $(u_1 - du_1/2, u_2, u_3)$. Adding these two, taking the limit $du_1 \to 0$, $du_2 \to 0$ and dividing by the area gives a contribution

$$\frac{1}{h_1h_2}\frac{\partial}{\partial u_1}(v_2h_2)$$

to the $\boldsymbol{e_3}$ component of $\boldsymbol{\nabla}\times\boldsymbol{v}$, and similarly the upper and lower sides of the rectangle generate a contribution

$$-\frac{1}{h_1h_2}\frac{\partial}{\partial u_2}(v_1h_1),$$

so the $\boldsymbol{e_3}$ component of $\boldsymbol{\nabla}\times\boldsymbol{v}$ is

$$\boldsymbol{e_3}\cdot\boldsymbol{\nabla}\times\boldsymbol{v} = \frac{1}{h_1h_2}\left(\frac{\partial}{\partial u_1}(v_2h_2) - \frac{\partial}{\partial u_2}(v_1h_1)\right).$$

The other components are found by permuting the indices. The determinant form of $\boldsymbol{\nabla}\times\boldsymbol{v}$ is

$$\boldsymbol{\nabla}\times\boldsymbol{v} = \frac{1}{h_1h_2h_3}\begin{vmatrix} h_1\boldsymbol{e_1} & h_2\boldsymbol{e_2} & h_3\boldsymbol{e_3} \\ \frac{\partial}{\partial u_1} & \frac{\partial}{\partial u_2} & \frac{\partial}{\partial u_3} \\ h_1v_1 & h_2v_2 & h_3v_3 \end{vmatrix}.$$

EXERCISES

6.1 Verify that for Cartesian coordinates the scale factors are all equal to 1.

6.2 A coordinate system (u, v, w) is related to Cartesian coordinates (x_1, x_2, x_3) by

$$x_1 = uvw, \quad x_2 = uv(1 - w^2)^{1/2}, \quad x_3 = (u^2 - v^2)/2.$$

(a) Find the scale factors h_u, h_v, h_w.
(b) Confirm that the (u, v, w) system is orthogonal.
(c) Find the volume element in the (u, v, w) system.

6.3 Find the scale factors and hence the volume element for the coordinate system (u, v, θ) defined by

$$x_1 = uv\cos\theta, \quad x_2 = uv\sin\theta, \quad x_3 = (u^2 - v^2)/2,$$

in which u and v are positive and $0 \le \theta < 2\pi$. Hence find the volume of the region enclosed by the curved surfaces $u = 1$ and $v = 1$.

6.4 Find the formula for ∇f in a general orthogonal curvilinear coordinate system by writing ∇f in Cartesian coordinates and then finding the component of ∇f in the $\boldsymbol{e}_1$ direction.

6.3 Cylindrical polar coordinates

Cylindrical polar coordinates (R, ϕ, z) are related to Cartesian coordinates (x_1, x_2, x_3) by

$$x_1 = R\cos\phi, \qquad x_2 = R\sin\phi, \qquad x_3 = z. \tag{6.10}$$

The transformation in the reverse direction is

$$R = \sqrt{x_1^2 + x_2^2}, \qquad \phi = \tan^{-1}(x_2/x_1), \qquad z = x_3. \tag{6.11}$$

The coordinate system is shown in Figure 6.5. The scale factors are found using (6.2):

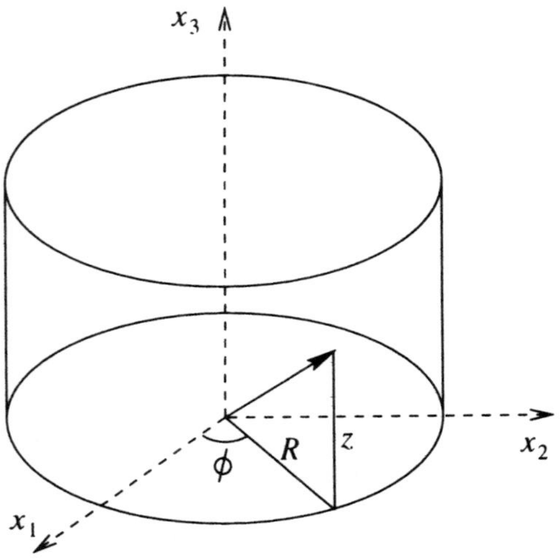

Fig. 6.5. Cylindrical polar coordinates.

$$
\begin{aligned}
h_R &= \left|\frac{\partial \boldsymbol{x}}{\partial R}\right| = |(\cos\phi, \sin\phi, 0)| = 1, \\
h_\phi &= \left|\frac{\partial \boldsymbol{x}}{\partial \phi}\right| = |(-R\sin\phi, R\cos\phi, 0)| = R, \\
h_z &= \left|\frac{\partial \boldsymbol{x}}{\partial z}\right| = |(0, 0, 1)| = 1.
\end{aligned}
$$

The unit vectors are found from (6.1):

$$
\begin{aligned}
\boldsymbol{e_R} &= \frac{\partial \boldsymbol{x}}{\partial R}\Big/ h_R = (\cos\phi, \sin\phi, 0), \\
\boldsymbol{e_\phi} &= \frac{\partial \boldsymbol{x}}{\partial \phi}\Big/ h_\phi = (-\sin\phi, \cos\phi, 0), \\
\boldsymbol{e_z} &= \frac{\partial \boldsymbol{x}}{\partial z}\Big/ h_z = (0, 0, 1).
\end{aligned}
$$

Note that the dot product of any two of the unit vectors is zero, so they obey the orthogonality condition. Also, since $\boldsymbol{e_R} \times \boldsymbol{e_\phi} = \boldsymbol{e_z}$, the coordinate system (R, ϕ, z) is right-handed.

Using the scale factors, area elements on each of the coordinate surfaces can be found from (6.6). For example on the cylindrical R coordinate surface,

$$dS = h_\phi h_z \, d\phi \, dz = R \, d\phi \, dz. \tag{6.12}$$

Similarly, the volume element is, from (6.7),

$$dV = h_R h_\phi h_z \, dR \, d\phi \, dz = R \, dR \, d\phi \, dz. \tag{6.13}$$

The formulae for the gradient, divergence, Laplacian and curl can be obtained by applying the results of Section 6.2:

$$\boldsymbol{\nabla} f = \frac{\partial f}{\partial R}\boldsymbol{e_R} + \frac{1}{R}\frac{\partial f}{\partial \phi}\boldsymbol{e_\phi} + \frac{\partial f}{\partial z}\boldsymbol{e_z}, \tag{6.14}$$

$$\boldsymbol{\nabla} \cdot \boldsymbol{v} = \frac{1}{R}\frac{\partial}{\partial R}(Rv_R) + \frac{1}{R}\frac{\partial v_\phi}{\partial \phi} + \frac{\partial v_z}{\partial z}, \tag{6.15}$$

$$\nabla^2 f = \frac{1}{R}\frac{\partial}{\partial R}\left(R\frac{\partial f}{\partial R}\right) + \frac{1}{R^2}\frac{\partial^2 f}{\partial \phi^2} + \frac{\partial^2 f}{\partial z^2}, \tag{6.16}$$

$$\begin{aligned} \boldsymbol{\nabla} \times \boldsymbol{v} &= \left(\frac{1}{R}\frac{\partial v_z}{\partial \phi} - \frac{\partial v_\phi}{\partial z}\right)\boldsymbol{e_R} + \left(\frac{\partial v_R}{\partial z} - \frac{\partial v_z}{\partial R}\right)\boldsymbol{e_\phi} \\ &\quad + \frac{1}{R}\left(\frac{\partial}{\partial R}(Rv_\phi) - \frac{\partial v_R}{\partial \phi}\right)\boldsymbol{e_z}. \end{aligned} \tag{6.17}$$

Example 6.3

Calculate the volume of a cone of radius a and height H.

In cylindrical polar coordinates the equation of the cone is $z = HR/a$ if the origin is chosen at the vertex of the cone. Then for a given value of z the range of R is $0 < R < az/H$. The volume is then

$$\begin{aligned} \iiint_V dV &= \int_0^H \int_0^{az/H} \int_0^{2\pi} R \, d\phi \, dR \, dz \\ &= \int_0^H \int_0^{az/H} 2\pi R \, dR \, dz \\ &= \int_0^H \pi \frac{a^2 z^2}{H^2} \, dz = \pi a^2 H/3. \end{aligned}$$

Example 6.4

Find the solution to Laplace's equation in cylindrical polar coordinates that only depends on the distance R from the axis.

If f obeys Laplace's equation $\nabla^2 f = 0$ and f depends only on R then from (6.16),

$$\frac{\partial}{\partial R}\left(R\frac{\partial f}{\partial R}\right) = 0, \qquad \text{so} \quad R\frac{\partial f}{\partial R} = c,$$

where c is a constant. Solving this differential equation gives

$$f = c \log R + d,$$

where d is another arbitrary constant.

6.4 Spherical polar coordinates

Spherical polar coordinates (r, θ, ϕ) are related to Cartesian coordinates by

$$x_1 = r \sin\theta \cos\phi, \qquad x_2 = r \sin\theta \sin\phi, \qquad x_3 = r\cos\theta, \tag{6.18}$$

and the inverse transformation is

$$r = \sqrt{x_1^2 + x_2^2 + x_3^2}, \quad \theta = \tan^{-1}\left(\frac{\sqrt{x_1^2 + x_2^2}}{x_3}\right), \quad \phi = \tan^{-1}\left(\frac{x_2}{x_1}\right). \tag{6.19}$$

Note that ϕ here is equivalent to the angle ϕ for cylindrical polar coordinates. However, the reader should be aware that different authors use different notation for the labelling of the angles in cylindrical and spherical polar coordinates. The variable r corresponds to its earlier use as the magnitude of the position vector $\boldsymbol{r}$, representing the distance from the origin.

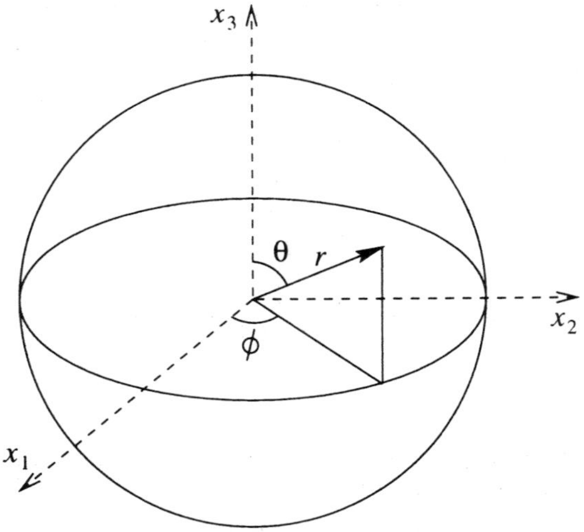

Fig. 6.6. Spherical polar coordinates.

A sketch of the coordinate system is shown in Figure 6.6. Note that the ranges of the three coordinates are

$$0 \le r < \infty, \qquad 0 \le \theta \le \pi, \qquad 0 \le \phi < 2\pi.$$

A slightly modified form of spherical polar coordinates is used to measure position on the surface of the Earth. The longitude is the angle ϕ and the latitude is $\pi/2 - \theta$.

The scale factors h_r, h_θ and h_ϕ are

$$\begin{aligned}
h_r &= |(\sin\theta\cos\phi, \sin\theta\sin\phi, \cos\theta)| = 1,\\
h_\theta &= |(r\cos\theta\cos\phi, r\cos\theta\sin\phi, -r\sin\theta)| = r,\\
h_\phi &= |(-r\sin\theta\sin\phi, r\sin\theta\cos\phi, 0)| = r\sin\theta.
\end{aligned}$$

The unit vectors are

$$\begin{aligned}
\boldsymbol{e_r} &= (\sin\theta\cos\phi, \sin\theta\sin\phi, \cos\theta),\\
\boldsymbol{e_\theta} &= (\cos\theta\cos\phi, \cos\theta\sin\phi, -\sin\theta),\\
\boldsymbol{e_\phi} &= (-\sin\phi, \cos\phi, 0)
\end{aligned}$$

and it can be verified by taking the dot product of pairs of the unit vectors that the system is orthogonal. The area element on the spherical r coordinate surface is

$$dS = h_\theta h_\phi\, d\theta\, d\phi = r^2\sin\theta\, d\theta\, d\phi \tag{6.20}$$

and the volume element is

$$dV = h_r h_\theta h_\phi\, dr\, d\theta\, d\phi = r^2\sin\theta\, dr\, d\theta\, d\phi. \tag{6.21}$$

The formulae for grad, div, Laplacian and curl are

$$\boldsymbol{\nabla} f = \frac{\partial f}{\partial r}\boldsymbol{e_r} + \frac{1}{r}\frac{\partial f}{\partial\theta}\boldsymbol{e_\theta} + \frac{1}{r\sin\theta}\frac{\partial f}{\partial\phi}\boldsymbol{e_\phi}, \tag{6.22}$$

$$\boldsymbol{\nabla}\cdot\boldsymbol{v} = \frac{1}{r^2}\frac{\partial}{\partial r}(r^2 v_r) + \frac{1}{r\sin\theta}\frac{\partial}{\partial\theta}(\sin\theta\, v_\theta) + \frac{1}{r\sin\theta}\frac{\partial v_\phi}{\partial\phi}, \tag{6.23}$$

$$\nabla^2 f = \frac{1}{r^2}\frac{\partial}{\partial r}\left(r^2\frac{\partial f}{\partial r}\right) + \frac{1}{r^2\sin\theta}\frac{\partial}{\partial\theta}\left(\sin\theta\frac{\partial f}{\partial\theta}\right) + \frac{1}{r^2\sin^2\theta}\frac{\partial^2 f}{\partial\phi^2}, \tag{6.24}$$

$$\begin{aligned}
\boldsymbol{\nabla}\times\boldsymbol{v} &= \frac{1}{r\sin\theta}\left(\frac{\partial}{\partial\theta}(\sin\theta\, v_\phi) - \frac{\partial v_\theta}{\partial\phi}\right)\boldsymbol{e_r}\\
&\quad + \frac{1}{r}\left(\frac{1}{\sin\theta}\frac{\partial v_r}{\partial\phi} - \frac{\partial}{\partial r}(r v_\phi)\right)\boldsymbol{e_\theta}\\
&\quad + \frac{1}{r}\left(\frac{\partial}{\partial r}(r v_\theta) - \frac{\partial v_r}{\partial\theta}\right)\boldsymbol{e_\phi}.
\end{aligned} \tag{6.25}$$

Example 6.5

Suppose a sphere of radius a has a variable density $\rho = \rho_0(1 - r/a)$ where ρ_0 is a constant. What is the total mass of the sphere?

The total mass M is just the volume integral of ρ. The limits on the integrals are $0 \leq r \leq a$, $0 \leq \theta \leq \pi$, $0 \leq \phi < 2\pi$ and the volume element is $dV = r^2 \sin\theta \, dr \, d\theta \, d\phi$. Choosing to do the integrals in the order r, θ, ϕ, the mass is

$$\begin{aligned}
M &= \iiint_V \rho_0 \, (1 - r/a) \, dV \\
&= \int_0^{2\pi} \int_0^{\pi} \int_0^{a} \rho_0 \, (1 - r/a) r^2 \sin\theta \, dr \, d\theta \, d\phi \\
&= \int_0^{2\pi} \int_0^{\pi} \rho_0 \left[r^3/3 - r^4/4a \right]_0^a \sin\theta \, d\theta \, d\phi \\
&= \int_0^{2\pi} \rho_0 \, a^3/12 \big[-\cos\theta \big]_0^{\pi} \, d\phi \\
&= \pi \rho_0 \, a^3/3.
\end{aligned}$$

Note that in triple integrals of this type, where the limits are constants, it is not strictly necessary to do the integrals one at a time. Thus it would have been possible to write after the second line,

$$M = \left[r^3/3 - r^4/4a \right]_0^a \big[-\cos\theta \big]_0^{\pi} \big[\phi \big]_0^{2\pi} = \pi \rho_0 \, a^3/3.$$

Example 6.6

What proportion of the Earth's surface lies further north than the 45° N latitude line?

The required region of the surface is $0 \leq \theta \leq \pi/4$, $0 \leq \phi < 2\pi$. Using the formula (6.20) for the area element, the area A is

$$A = \iint dS = \int_0^{2\pi} \int_0^{\pi/4} r^2 \sin\theta \, d\theta \, d\phi = 2\pi r^2 \, (1 - \cos(\pi/4)) = \pi r^2 (2 - \sqrt{2}).$$

As a proportion of the total area this is $A/4\pi r^2 = (2 - \sqrt{2})/4 \approx 0.15$.

Summary of Chapter 6

- A curvilinear coordinate system (u_1, u_2, u_3) is related to a Cartesian coordinate system (x_1, x_2, x_3) by $x_i = x_i(u_1, u_2, u_3)$.
- The *unit vectors* $\boldsymbol{e_i}$ and the *scale factors* h_i are defined for $i = 1$ by

$$h_1 = \left|\frac{\partial \boldsymbol{x}}{\partial u_1}\right|, \qquad \boldsymbol{e_1} = \frac{\partial \boldsymbol{x}}{\partial u_1}\Big/ h_1,$$

  and similarly for $i = 2, 3$.
- The system (u_1, u_2, u_3) is *orthogonal* if $\boldsymbol{e_i} \cdot \boldsymbol{e_j} = \delta_{ij}$.
- The volume element in the (u_1, u_2, u_3) system is $dV = h_1 h_2 h_3\, du_1\, du_2\, du_3$.
- Formulae for grad, div and curl in the (u_1, u_2, u_3) system can be written down in terms of the scale factors and the unit vectors.
- The two most important curvilinear coordinate systems are cylindrical polar coordinates,

$$x_1 = R\cos\phi, \qquad x_2 = R\sin\phi, \qquad x_3 = z$$

  and spherical polar coordinates,

$$x_1 = r\sin\theta\cos\phi, \qquad x_2 = r\sin\theta\sin\phi, \qquad x_3 = r\cos\theta.$$

EXERCISES

6.5 A cylindrical apple corer of radius a cuts through a spherical apple of radius b. How much of the apple does it remove?

6.6 Find the proportion of the Earth's volume that is less than 30° away from the Equator.

6.7 Find the divergence and curl of the unit vector $\boldsymbol{e_\phi}$ in spherical polar coordinates.

6.8 Find $\boldsymbol{u} \cdot \boldsymbol{\nabla u}$ for the vector $\boldsymbol{u} = \boldsymbol{e_\phi}$ in cylindrical polar coordinates.

6.9 Find a formula for the R component of the Laplacian of a vector field, $\nabla^2 \boldsymbol{v}$, in cylindrical polar coordinates. Verify that the components of the Laplacian of $\boldsymbol{v}$ are not equal to the Laplacians of the components of $\boldsymbol{v}$.

7
Cartesian Tensors

7.1 Coordinate transformations

At the very beginning of this book vectors and scalars were defined as 'physical quantities'. But what does this mean mathematically? In this chapter a precise mathematical statement is developed, using the idea that the physical quantity exists independently of any coordinate system that may be used. This new mathematical definition of vectors and scalars is generalised to define a wider class of objects known as tensors. Throughout this chapter attention is restricted to Cartesian coordinate systems.

Consider a rotation of a two-dimensional Cartesian coordinate system x_1, x_2 through an angle θ (Figure 7.1) to give a new coordinate system x_1', x_2'. Then by carrying out some simple geometrical constructions it can be seen that the coordinates of a point P in the x_1, x_2 system are related to those in the x_1', x_2' system by the equations

$$x_1' = x_1 \cos\theta + x_2 \sin\theta, \tag{7.1}$$
$$x_2' = x_2 \cos\theta - x_1 \sin\theta, \tag{7.2}$$

or in matrix form,

$$\begin{pmatrix} x_1' \\ x_2' \end{pmatrix} = \begin{pmatrix} \cos\theta & \sin\theta \\ -\sin\theta & \cos\theta \end{pmatrix} \begin{pmatrix} x_1 \\ x_2 \end{pmatrix}.$$

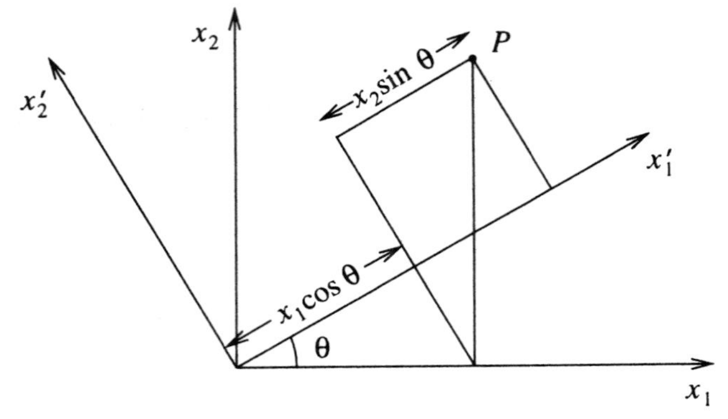

Fig. 7.1. Rotation of Cartesian coordinates through an angle θ.

The 2×2 matrix relating (x_1', x_2') to (x_1, x_2) will be referred to as L:

$$L = \begin{pmatrix} \cos\theta & \sin\theta \\ -\sin\theta & \cos\theta \end{pmatrix}. \tag{7.3}$$

The matrix multiplication can be written in suffix notation, since

$$\begin{aligned} x_1' &= L_{11}x_1 + L_{12}x_2 = L_{1j}x_j, \\ x_2' &= L_{21}x_1 + L_{22}x_2 = L_{2j}x_j, \end{aligned}$$

where the repeated suffix j implies summation, so

$$x_i' = L_{ij}x_j. \tag{7.4}$$

The rotation matrix L_{ij} has one particularly important property. The inverse of the matrix is a rotation through $-\theta$,

$$L^{-1} = \begin{pmatrix} \cos(-\theta) & \sin(-\theta) \\ -\sin(-\theta) & \cos(-\theta) \end{pmatrix} = \begin{pmatrix} \cos\theta & -\sin\theta \\ \sin\theta & \cos\theta \end{pmatrix},$$

which is the transpose of the matrix L. Thus $LL^T = I$, or in suffix notation, $L_{ij}L^T_{jk} = \delta_{ik}$. Since $L^T_{jk} = L_{kj}$, this can be written

$$L_{ij}L_{kj} = \delta_{ik}. \tag{7.5}$$

A matrix with this property, that its inverse is equal to its transpose, is said to be *orthogonal.* Using this property, the inverse of the transformation can be written down, simply by transposing the suffices:

$$x_i = L_{ji}x_j'. \tag{7.6}$$

Another important property of the matrix L is that its determinant is

$$|L| = \cos^2\theta + \sin^2\theta = 1.$$

So far we have only considered a two-dimensional rotation of coordinates. Consider now a general three-dimensional rotation. For a position vector $\boldsymbol{x} = x_1\boldsymbol{e}_1 + x_2\boldsymbol{e}_2 + x_3\boldsymbol{e}_3$, the i component in the dashed frame is defined by

$$x'_i = \boldsymbol{e}'_i \cdot \boldsymbol{x} = \boldsymbol{e}'_i \cdot \boldsymbol{e}_1 x_1 + \boldsymbol{e}'_i \cdot \boldsymbol{e}_2 x_2 + \boldsymbol{e}'_i \cdot \boldsymbol{e}_3 x_3 = \boldsymbol{e}'_i \cdot \boldsymbol{e}_j x_j.$$

This is of the form (7.4), where

$$L_{ij} = \boldsymbol{e}'_i \cdot \boldsymbol{e}_j, \tag{7.7}$$

so L_{ij} is the cosine of the angle between $\boldsymbol{e}'_i$ and $\boldsymbol{e}_j$. By the same argument, the matrix which transforms from the dashed frame to the undashed frame has i,j element $\boldsymbol{e}_i \cdot \boldsymbol{e}'_j = L_{ji}$, so again we see that the inverse of L is its transpose. Since $LL^T = I$, the determinant of L obeys $|L|^2 = 1$, so $|L| = \pm 1$. Orthogonal matrices with $|L| = 1$ represent rotations, while those with $|L| = -1$ are reflections.

From (7.4) and (7.6), two further important properties of L follow:

$$\frac{\partial x'_i}{\partial x_j} = L_{ij} \qquad \text{and} \qquad \frac{\partial x_i}{\partial x'_j} = L_{ji}. \tag{7.8}$$

7.2 Vectors and scalars

Now consider a vector $\boldsymbol{v}$. Its components transform from one coordinate system to another in the same way as the coordinates of a point, so

$$v'_i = L_{ij} v_j. \tag{7.9}$$

This equation gives the mathematical definition of a vector: $\boldsymbol{v}$ is a vector if its components transform according to the rule (7.9) under a rotation of the coordinate axes.

Similarly, a scalar s is defined by the property that its value is unchanged by a rotation of coordinates, so

$$s' = s. \tag{7.10}$$

Using these new definitions of scalars and vectors, in terms of their transformation properties under a rotation of coordinate axes, a number of rigorous results can be proved, as illustrated in the following examples. Suffix notation and the summation convention are used throughout.

Example 7.1

Suppose that $\boldsymbol{a}$ and $\boldsymbol{b}$ are vectors. Show that their dot product $\boldsymbol{a} \cdot \boldsymbol{b}$ is a scalar.

Since $\boldsymbol{a}$ and $\boldsymbol{b}$ are vectors, their components transform under rotation according to

$$a_i' = L_{ij} a_j, \qquad b_i' = L_{ij} b_j.$$

Now to show $\boldsymbol{a} \cdot \boldsymbol{b}$ is a scalar, we must show that its value in the dashed frame is the same as its value in the undashed frame.

$$(\boldsymbol{a} \cdot \boldsymbol{b})' = a_i' b_i' = L_{ij} a_j L_{ik} b_k = L_{ij} L_{ik} a_j b_k \tag{7.11}$$

$$= \delta_{jk} a_j b_k = a_k b_k = \boldsymbol{a} \cdot \boldsymbol{b}, \tag{7.12}$$

so $\boldsymbol{a} \cdot \boldsymbol{b}$ is a scalar.

Example 7.2

Suppose that f is a scalar field. Show that $\boldsymbol{\nabla} f$ is a vector.

If f is a scalar then $f = f'$. To show that $\boldsymbol{\nabla} f$ is a vector we need to determine how it transforms under a rotation of coordinates.

$$(\boldsymbol{\nabla} f)_i' = \frac{\partial f'}{\partial x_i'} = \frac{\partial f}{\partial x_i'} = \frac{\partial f}{\partial x_j} \frac{\partial x_j}{\partial x_i'}$$

using the chain rule. Now making use of (7.8),

$$\frac{\partial f}{\partial x_i'} = L_{ij} \frac{\partial f}{\partial x_j},$$

so $\boldsymbol{\nabla} f$ obeys the transformation rule for a vector.

Example 7.3

A quantity is defined in a two-dimensional Cartesian coordinate system by $\boldsymbol{u} = (a x_2, b x_1)^T$. Show that this quantity can only be a vector if $a = -b$.

If $\boldsymbol{u}$ is a vector, it must transform according to the rule $u_i' = L_{ij} u_j$ where L_{ij} is the 2×2 rotation matrix (7.3). This gives

$$\boldsymbol{u}' = \begin{pmatrix} a x_2 \cos\theta + b x_1 \sin\theta \\ -a x_2 \sin\theta + b x_1 \cos\theta \end{pmatrix},$$

but from the definition of $\boldsymbol{u}$ we also have

$$\boldsymbol{u}' = \begin{pmatrix} a x_2' \\ b x_1' \end{pmatrix} = \begin{pmatrix} -a x_1 \sin\theta + a x_2 \cos\theta \\ b x_1 \cos\theta + b x_2 \sin\theta \end{pmatrix}.$$

By comparing these two expressions we can see that they only agree if $a = -b$, so this is the condition for $\boldsymbol{u}$ to be a vector.

7.3 Tensors

The definition of a vector as a quantity which transforms in a certain way under a rotation of coordinates can be extended to define a more general class of objects called tensors, which may have more than one free suffix. A quantity is a *tensor* if each of the free suffices transforms according to the rule (7.4). For example, consider a quantity T_{ij} that has two free suffices. This quantity is a tensor if its components in the dashed frame are related to those in the undashed frame by the equation

$$T'_{ij} = L_{ik}L_{jm}T_{km}. \tag{7.13}$$

The *rank* or *order* of the tensor is the number of free suffices, so the quantity T_{ij} obeying (7.13) is said to be a second-rank tensor. A tensor may have any number of free suffices. For example, a third-rank tensor P_{ijk} transforms according to the rule

$$P'_{ijk} = L_{ip}L_{jq}L_{kr}P_{pqr}. \tag{7.14}$$

The rule for a tensor of rank one is the same as the rule for a vector, so a vector can be regarded as tensor of rank one. Similarly, a scalar can be thought of as a tensor of rank zero.

We have already met one second-rank tensor, δ_{ij}, and a third-rank tensor, ϵ_{ijk}. Tensors can also be constructed from vectors, for example $\partial u_i/\partial x_j$ is a tensor. The demonstration that these quantities are indeed tensors is given in the following examples.

Example 7.4

Show that δ_{ij} is a tensor.

Consider the quantity $L_{ik}L_{jm}\delta_{km}$. From the substitution property of δ_{ij}, this is $L_{ik}L_{jk}$, which from the property (7.5) of L is δ_{ij}. Now $\delta'_{ij} = \delta_{ij}$, since δ_{ij} is defined the same way in any coordinate system. Thus δ_{ij} obeys the tensor transformation law, $\delta'_{ij} = L_{ik}L_{jm}\delta_{km}$.

Example 7.5

Show that ϵ_{ijk} is a tensor.

Since ϵ_{ijk} has three suffices, the appropriate transformation to consider is $L_{ip}L_{jq}L_{kr}\epsilon_{pqr}$. Using (4.10), this is $\epsilon_{ijk}|L| = \epsilon_{ijk}$, since $|L| = 1$ for a rotation. As for δ_{ij}, ϵ_{ijk} is defined in the same way in all coordinate systems so $\epsilon'_{ijk} = \epsilon_{ijk} = L_{ip}L_{jq}L_{kr}\epsilon_{pqr}$. Therefore ϵ_{ijk} is a third-rank tensor.

Example 7.6

If $\boldsymbol{u}$ is a vector, show that $\partial u_i/\partial x_j$ is a second-rank tensor.

Since $\boldsymbol{u}$ is a vector, $u'_i = L_{ik}u_k$.

$$\frac{\partial u'_i}{\partial x'_j} = L_{ik}\frac{\partial u_k}{\partial x'_j} = L_{ik}\frac{\partial u_k}{\partial x_l}\frac{\partial x_l}{\partial x'_j} = L_{ik}L_{jl}\frac{\partial u_k}{\partial x_l},$$

which is the transformation rule for a second-rank tensor.

7.3.1 The quotient rule

Tensors often appear as quantities relating two vectors, for example

$$a_i = T_{ij}b_j. \tag{7.15}$$

The quotient rule states that if (7.15) holds in all coordinate systems and for any vector $\boldsymbol{b}$ the resulting quantity $\boldsymbol{a}$ is a vector, then T_{ij} is a tensor.

Proof

The quotient rule is proved as follows: Since $\boldsymbol{a}$ is a vector,

$$a'_i = L_{ik}a_k = L_{ik}T_{kj}b_j.$$

Since $\boldsymbol{b}$ is a vector, it obeys $b_j = L_{mj}b'_m$ (note that this is the inverse transformation, from the dashed to the undashed frame, so the suffices of L are transposed). Substituting for b_j gives

$$a'_i = L_{ik}T_{kj}L_{mj}b'_m.$$

But since (7.15) holds in all coordinate systems,

$$a'_i = T'_{im}b'_m.$$

Subtracting these two results,

$$(T'_{im} - L_{ik}T_{kj}L_{mj})b'_m = 0.$$

If this result holds for any vector $\boldsymbol{b}$, then the quantity in brackets must be zero, so

$$T'_{im} = L_{ik}L_{mj}T_{kj}.$$

Therefore, T_{ij} is a second-rank tensor. □

A more general form of the quotient rule also holds: if an mth rank tensor a is linearly related to an nth rank tensor b through a quantity T with $m+n$ suffices, then T is a tensor of rank $m+n$.

EXERCISES

7.1 Show that the definition $L_{ij} = \boldsymbol{e}'_i \cdot \boldsymbol{e}_j$ is consistent with the matrix given in (7.3).

7.2 If $\boldsymbol{u}$ is a vector field, show that $\boldsymbol{\nabla} \cdot \boldsymbol{u}$ is a scalar field.

7.3 Given that $\boldsymbol{a}$ and $\boldsymbol{b}$ are vectors, show that the quantity $a_i b_j$ is a second-rank tensor.

7.4 Show that in a two-dimensional Cartesian coordinate system (x_1, x_2) the quantity

$$T_{ij} = \begin{pmatrix} x_1 x_2 & -x_1^2 \\ x_2^2 & -x_1 x_2 \end{pmatrix}$$

is a tensor.

7.5 If ϕ is a scalar field, show that the quantity

$$T_{jk} = \frac{\partial^2 \phi}{\partial x_j \partial x_k}$$

is a second-rank tensor.

7.6 If T_{ij} is a tensor, show that T_{ii} is a scalar.

7.7 Write the divergence theorem in the form of suffix notation and hence obtain the analogue of the divergence theorem for a second-rank tensor T_{ij}:

$$\iiint_V \frac{\partial T_{ij}}{\partial x_j} \, dV = \oiint_S T_{ij} n_j \, dS. \tag{7.16}$$

7.8 Write down the transformation rule for a tensor of rank four.

7.9 If Q_{ijkl} is a tensor of rank four, show that Q_{ijjl} is a tensor of rank two.

7.10 A quantity u_i has the property that for any vector $\boldsymbol{a}$, $u_i a_i$ is a scalar. Show that the u_i are the components of a vector. (This is a form of the quotient rule.)

7.3.2 Symmetric and anti-symmetric tensors

A second-rank tensor T_{ij} is said to be *symmetric* if $T_{ij} = T_{ji}$ and *anti-symmetric* if $T_{ij} = -T_{ji}$. A tensor of rank greater than two can be symmetric or anti-symmetric with respect to any pair of indices. For example δ_{ij} is a symmetric tensor, while ϵ_{ijk} is anti-symmetric with respect to any two of its indices.

It is important to verify that symmetry is a physical property of tensors, i.e. that if a tensor is symmetric in a Cartesian coordinate system it is also symmetric in other Cartesian coordinate systems. This can be confirmed as follows: suppose that A_{ij} is a symmetric tensor, so $A_{ij} = A_{ji}$. Then in a rotated frame,

$$A'_{ij} = L_{ik}L_{jm}A_{km} = L_{jm}L_{ik}A_{mk} = A'_{ji},$$

so A'_{ij} is also symmetric.

Example 7.7

Show that any second-rank tensor T_{ij} can be written as the sum of a symmetric tensor and an anti-symmetric tensor.

For any tensor T_{ij}, the tensor $S_{ij} = T_{ij} + T_{ji}$ is symmetric. Similarly, $A_{ij} = T_{ij} - T_{ji}$ is anti-symmetric. Since $S_{ij} + A_{ij} = 2T_{ij}$, T_{ij} can be written as $T_{ij} = S_{ij}/2 + A_{ij}/2$.

Example 7.8

The second-rank tensor T_{ij} obeys $\epsilon_{ijk}T_{jk} = 0$. Show that T_{ij} is a symmetric tensor.

By expanding out the implied double sum, for $i = 1$ we have $\epsilon_{123}T_{23} + \epsilon_{132}T_{32} = 0$, which gives $T_{23} = T_{32}$. Similarly the other required results follow from taking $i = 2$ and $i = 3$.

The same result may be obtained more elegantly by multiplying the given equation $\epsilon_{ijk}T_{jk} = 0$ by ϵ_{mni}:

$$\begin{aligned} 0 &= \epsilon_{mni}\epsilon_{ijk}T_{jk} \\ &= (\delta_{mj}\delta_{nk} - \delta_{mk}\delta_{nj})T_{jk} \\ &= T_{mn} - T_{nm}, \end{aligned}$$

so $T_{mn} = T_{nm}$.

7.3.3 Isotropic tensors

The two tensors δ_{ij} and ϵ_{ijk} have a special property. Their components are the same in all coordinate systems. A tensor with this property is said to be *isotropic*. Isotropic tensors are of great importance physically, and it turns out that there are very few examples of isotropic tensors. This is illustrated by the following results.

Theorem 7.1

There are no non-trivial isotropic first-rank tensors.

Proof

Suppose that there exists an isotropic first-rank tensor (i.e. an isotropic vector), $\boldsymbol{u} = (u_1, u_2, u_3)$. Now consider a rotation through $\pi/2$ about the x_3-axis, which is given by the matrix

$$L = \begin{pmatrix} 0 & 1 & 0 \\ -1 & 0 & 0 \\ 0 & 0 & 1 \end{pmatrix}. \tag{7.17}$$

If $\boldsymbol{u}$ is a first-rank tensor then $u'_i = L_{ij}u_j = (u_2, -u_1, u_3)$. Now if $\boldsymbol{u}$ is isotropic, $u'_i = u_i$, so $u_1 = u_2$ and $u_2 = -u_1$. Therefore $u_1 = u_2 = 0$. By considering a rotation about the x_1-axis in a similar way, it can be shown also that $u_3 = 0$, so the only solution is $\boldsymbol{u} = (0, 0, 0)$. □

Theorem 7.2

The most general isotropic second-rank tensor is a multiple of δ_{ij}.

Proof

Suppose that a_{ij} is an isotropic second-rank tensor. Consider the rotation through $\pi/2$ about the x_3-axis given by (7.17). a_{ij} must obey $a'_{ij} = L_{im}L_{jn}a_{mn}$, which in terms of matrix multiplication is $a' = LaL^T$. Carrying out these matrix multiplications gives the result

$$LaL^T = \begin{pmatrix} a_{22} & -a_{21} & a_{23} \\ -a_{12} & a_{11} & -a_{13} \\ a_{32} & -a_{31} & a_{33} \end{pmatrix}. \tag{7.18}$$

This must be equal to a_{ij} if the tensor a_{ij} is isotropic. The terms on the diagonal give $a_{11} = a_{22}$. The other terms give $a_{13} = a_{23}$ and $a_{23} = -a_{13}$, from which $a_{13} = a_{23} = 0$. By considering the analogous rotations about the other coordinate axes it follows that $a_{11} = a_{22} = a_{33}$ and that all the off-diagonal terms are zero, so $a_{ij} = \lambda\delta_{ij}$, where λ is an arbitrary constant. □

Theorem 7.3

The most general isotropic third-rank tensor is a multiple of ϵ_{ijk}.

Proof

If a_{ijk} is an isotropic third-rank tensor, then

$$a_{ijk} = a'_{ijk} = L_{ip}L_{jq}L_{kr}a_{pqr}. \tag{7.19}$$

Consider the same rotation (7.17), for which the only non-zero elements of L are $L_{12} = 1$, $L_{21} = -1$ and $L_{33} = 1$. Therefore for any choice of i, j and k in (7.19), only one term on the r.h.s. is non-zero. Choosing $(i, j, k) = (1, 1, 1)$ gives $a_{111} = a_{222}$ and the choice $(i, j, k) = (2, 2, 2)$ gives $a_{222} = -a_{111}$, so $a_{111} = a_{222} = 0$. A different choice of rotation matrix would yield $a_{333} = 0$.

By making further choices of (i, j, k) the following equations can be obtained: $a_{112} = -a_{221}$, $a_{221} = a_{112}$, $a_{122} = a_{211}$, $a_{211} = -a_{122}$, $a_{121} = -a_{212}$, $a_{212} = a_{121}$. From these and the analogous equations involving the suffices 2 and 3 it follows that all 18 elements with two suffices equal are zero.

Finally, by considering the cases when i, j and k are all different, (7.19) gives $a_{123} = -a_{213}$, $a_{231} = -a_{132}$, $a_{312} = -a_{321}$. The analogous equations for rotations about the other axes can be used to show that $a_{123} = a_{231} = a_{312} = -a_{321} = -a_{132} = -a_{213}$, so that $a_{ijk} = \lambda\epsilon_{ijk}$ for some constant λ. □

Theorem 7.4

The most general isotropic fourth-rank tensor is

$$a_{ijkl} = \lambda\delta_{ij}\delta_{kl} + \mu\delta_{ik}\delta_{jl} + \nu\delta_{il}\delta_{jk}, \tag{7.20}$$

where λ, μ and ν are constants.

Proof

An isotropic fourth-rank tensor must obey

$$a_{ijkl} = L_{ip}L_{jq}L_{kr}L_{ls}a_{pqrs}. \tag{7.21}$$

Using the rotation (7.17), only one of the 81 terms in the implied sum on the r.h.s. is non-zero. Since $L_{12} = 1$, $L_{21} = -1$ and $L_{33} = 1$, a suffix 1 on the l.h.s. becomes a suffix 2 on the r.h.s., a suffix 2 on the l.h.s. becomes a suffix 1 on the r.h.s. and changes the sign, while a suffix 3 remains unchanged. By applying these rules, $a_{1113} = a_{2223} = -a_{1113}$, so $a_{1113} = a_{2223} = 0$. Similarly, any other term with three suffices equal and the fourth one different must be zero. Also $a_{2113} = -a_{1223} = -a_{2113}$ so $a_{2113} = a_{1223} = 0$ and all similar terms with only one pair of equal suffices are zero.

The only remaining terms are those with two pairs of equal suffices and those with all four suffices equal. Applying the rotation (7.17) to terms in which the first two suffices are equal and the last two suffices are equal gives $a_{1122} = a_{2211}$, $a_{1133} = a_{2233}$ and $a_{3322} = a_{3311}$. Using the rotations about the other coordinate axes it follows that these six terms are all equal. Similarly, $a_{1212} = a_{2121} = a_{1313} = a_{2323} = a_{3131} = a_{3232}$ and $a_{1221} = a_{2112} = a_{1331} = a_{2332} = a_{3113} = a_{3223}$. The terms with all four suffices equal must obey $a_{1111} = a_{2222} = a_{3333}$. Thus there can be at most four independent components of the tensor, a_{1122}, a_{1212}, a_{1221} and a_{1111}.

To proceed it is necessary to consider a different rotation, for example the rotation through an arbitrary angle θ about the x_3-axis given by

$$L = \begin{pmatrix} \cos\theta & \sin\theta & 0 \\ -\sin\theta & \cos\theta & 0 \\ 0 & 0 & 1 \end{pmatrix}. \tag{7.22}$$

Using this rotation, a_{1111} is related to all the terms with suffices equal to 1 or 2. Applying (7.21) gives

$$\begin{aligned} a_{1111} &= \cos^4\theta\, a_{1111} + \sin^4\theta\, a_{2222} \\ &\quad + \sin^2\theta\cos^2\theta(a_{1122} + a_{2211} + a_{1212} + a_{2121} + a_{1221} + a_{2112}). \end{aligned}$$

Simplifying this equation and using the relations above, the trigonometric factors cancel out leaving

$$a_{1111} = a_{1122} + a_{1212} + a_{1221}, \tag{7.23}$$

so in fact there are only three independent components, which can be labelled $a_{1122} = \lambda$, $a_{1212} = \mu$, $a_{1221} = \nu$. The tensor a_{ijkl} can therefore be written in terms of λ, μ and ν in the form (7.20). Note that this ensures that (7.23) is satisfied.

□

7.4 Physical examples of tensors

Tensors appear in many contexts, including fluid mechanics, solid mechanics and general relativity. Some of these applications will be described in Chapter 8. The following two sections briefly consider two other examples of tensors.

7.4.1 Ohm's law

Ohm's law states that there is a linear relationship between the electric current $\boldsymbol{j}$ flowing through a material and the electric field $\boldsymbol{E}$ applied to the material. This can be written

$$\boldsymbol{j} = \sigma \boldsymbol{E}, \tag{7.24}$$

where the constant of proportionality σ is known as the conductivity (an inverse measure of electrical resistance). Note that (7.24) forces the vectors $\boldsymbol{j}$ and $\boldsymbol{E}$ to be parallel. For some materials, this may be true, but consider a substance with a layered structure made of different materials (Figure 7.2). For this material,

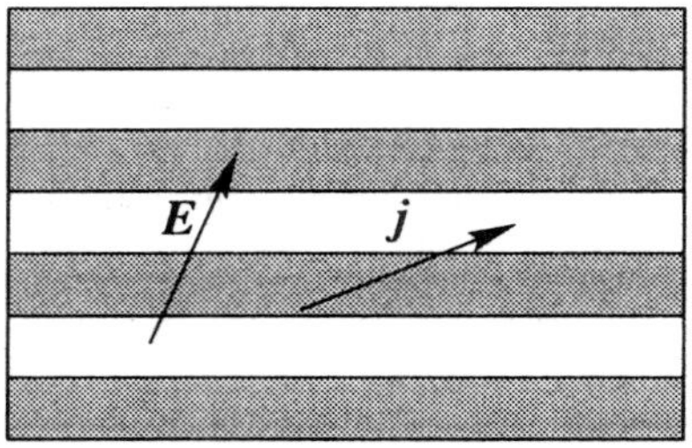

Fig. 7.2. For a material made up of layers, the electric field $\boldsymbol{E}$ and the electric current $\boldsymbol{j}$ may not be parallel.

current may flow more easily along the layers than across them. For example, if the substance is made of alternate layers of a conductor and an insulator, then current can only flow along the layers, regardless of the direction of the electric field.

It is useful therefore to have an alternative to (7.24) in which $\boldsymbol{j}$ and $\boldsymbol{E}$ do not have to be parallel. This can be achieved by introducing the *conductivity tensor*, σ_{ik}, which relates $\boldsymbol{j}$ and $\boldsymbol{E}$ through the equation

$$j_i = \sigma_{ik} E_k. \tag{7.25}$$

Since $\boldsymbol{j}$ and $\boldsymbol{E}$ are vectors, it follows from the quotient rule that σ_{ik} is a tensor.

The values of σ_{ik} depend on the properties of the material. For example, suppose that there are alternating layers of a conductor with conductivity σ_0 and an insulator. If axes are chosen such that the x_3 direction is perpendicular to the layers, then in this coordinate system

$$\sigma_{ik} = \begin{pmatrix} \sigma_0 & 0 & 0 \\ 0 & \sigma_0 & 0 \\ 0 & 0 & 0 \end{pmatrix}.$$

Now suppose that the material has no such layered structure, so that there is no preferred direction and is made of a uniform material with conductivity σ_0. Such a material is said to be *isotropic*, meaning 'the same in all directions'. In this case $\sigma_{ik} = \sigma_0 \, \delta_{ik}$, so

$$j_i = \sigma_{ik} E_k = \sigma_0 \, \delta_{ik} E_k = \sigma_0 \, E_i$$

and so the simple rule

$$\boldsymbol{j} = \sigma_0 \, \boldsymbol{E}$$

holds. This is why δ_{ik} is said to be an isotropic tensor: it represents the relationship between two vectors that are always parallel, regardless of their direction.

7.4.2 The inertia tensor

Consider a body rotating with angular velocity $\boldsymbol{\Omega}$. Then, as shown in Section 1.3.1, the velocity vector at the position vector $\boldsymbol{r}$ is

$$\boldsymbol{v} = \boldsymbol{\Omega} \times \boldsymbol{r}.$$

The angular momentum of a particle of mass m is $\boldsymbol{h} = m\boldsymbol{r} \times \boldsymbol{v}$. The total angular momentum of a rotating body can then be determined as a volume integral, by considering dividing the body into small volume elements dV each with mass $\rho \, dV$, where ρ is the density of the body. The total angular momentum $\boldsymbol{H}$ is therefore given by

$$\begin{aligned} H_i &= \iiint_V \rho \, (\boldsymbol{r} \times \boldsymbol{v})_i \, dV \\ &= \iiint_V \rho \, (\boldsymbol{r} \times (\boldsymbol{\Omega} \times \boldsymbol{r}))_i \, dV \\ &= \iiint_V \rho \left(r^2 \Omega_i - (\boldsymbol{r} \cdot \boldsymbol{\Omega}) r_i \right) dV \end{aligned}$$

$$= \iiint_V \rho \left(r^2 \delta_{ij} \Omega_j - r_j \Omega_j r_i\right) \, dV$$
$$= \iiint_V \rho \left(r^2 \delta_{ij} - r_i r_j\right) \Omega_j \, dV.$$

Since Ω_j is a constant it can be taken out of the integral, leaving the equation

$$H_i = I_{ij} \Omega_j, \qquad (7.26)$$

where I_{ij} is called the *inertia tensor* of the body and is defined by

$$I_{ij} = \iiint_V \rho \left(r^2 \delta_{ij} - r_i r_j\right) \, dV. \qquad (7.27)$$

Note that as in the previous example, the tensor appears as a quantity relating two vectors, and the quotient rule confirms that I_{ij} is a tensor. The inertia tensor is an example of a symmetric tensor, since it is clear that $I_{ij} = I_{ji}$.

Example 7.9

Find the inertia tensor for a cube with sides of length $2a$ and constant density ρ, for rotations about its centre.

To find I_{ij} we need to compute two volume integrals. First,

$$\begin{aligned}
\iiint_V \rho r^2 \, dV &= \int_{-a}^{a} \int_{-a}^{a} \int_{-a}^{a} \rho \left(x^2 + y^2 + z^2\right) \, dx \, dy \, dz \\
&= 3\rho \int_{-a}^{a} \int_{-a}^{a} \int_{-a}^{a} x^2 \, dx \, dy \, dz \\
&= 3\rho \, (2a)(2a) \int_{-a}^{a} x^2 \, dx \\
&= 8\rho \, a^5 = M a^2,
\end{aligned}$$

where $M = 8\rho \, a^3$ is the mass of the cube. The second volume integral is

$$\iiint_V \rho \, r_i r_j \, dV.$$

For $i \neq j$ this is zero, since for example the integral of xy is zero since this is an odd function of x and y. For $i = j$, for example $i = j = 1$, we have

$$\iiint_V \rho \, x^2 \, dV = M a^2 / 3$$

from the working of the first integral. Putting the two parts together,

$$I_{ij} = M a^2 \delta_{ij} - M a^2 \delta_{ij} / 3 = \frac{2}{3} M a^2 \delta_{ij}.$$

Note that the inertia tensor is isotropic. This means that for a cube rotating about its centre, the rotation vector and angular momentum vector are always parallel.

Summary of Chapter 7

- Under a rotation of coordinate axes from a frame with unit vectors $\boldsymbol{e}_i$ to a frame with unit vectors $\boldsymbol{e}'_i$, the coordinates of a point are related by

$$x'_i = L_{ij}x_j$$

where $L_{ij} = \boldsymbol{e}'_i \cdot \boldsymbol{e}_j$.
- The inverse of the transformation is

$$x_i = L_{ji}x'_j,$$

so $L^{-1} = L^T$. Such a matrix is said to be *orthogonal.* In suffix notation, this result is written

$$L_{ij}L_{kj} = \delta_{ik}.$$

- A *scalar* s has the same value in each frame, $s' = s$.
- A *vector* v transforms according to the rule $v'_i = L_{ij}v_j$.
- If a quantity T_{ij} transforms according to the rule $T'_{ij} = L_{ik}L_{jm}T_{km}$ then T_{ij} is a *tensor* of second rank. The *rank* of a tensor is the number of free suffices. Thus vectors are tensors of rank one and scalars are tensors of rank zero.
- The quotient rule says that if $a_i = T_{ij}b_j$ and $\boldsymbol{a}$ is a vector for any choice of the vector $\boldsymbol{b}$, then T_{ij} is a tensor.
- A tensor T_{ij} is symmetric if $T_{ij} = T_{ji}$ and anti-symmetric if $T_{ij} = -T_{ji}$.
- δ_{ij} and ϵ_{ijk} are tensors of a special type known as isotropic tensors. This means that their components do not change when the coordinate axes are rotated. A second-rank isotropic tensor must be a multiple of δ_{ij} and a third-rank isotropic tensor must be a multiple of ϵ_{ijk}.
- In physical systems, tensors frequently arise as quantities relating two vectors. This allows two vectors to be linearly related to each other without being parallel. Examples include the conductivity tensor and the inertia tensor.

EXERCISES

7.11 B_{rs} is an anti-symmetric tensor, so $B_{rs} = -B_{sr}$. Show that the anti-symmetry persists in a rotated frame, i.e. $B'_{rs} = -B'_{sr}$.

7.12 If B_{rs} is an anti-symmetric tensor, show that $B_{rr} = 0$.

7.13 The third-rank tensor A_{ijk} is symmetric with respect to its first two suffices but anti-symmetric with respect to the second and third suffices. Show that all elements of A_{ijk} must be zero.

7.14 A quantity A_{ij} is related to a vector $\boldsymbol{B}$ by $A_{ij} = \epsilon_{ijk} B_k$.
(a) Show that A_{ij} is a tensor and describe its symmetry property.
(b) Find an equation for $\boldsymbol{B}$ in terms of A_{ij}.

7.15 Find an isotropic fourth-rank tensor that can be written in terms of ϵ_{ijk}.

7.16 Write down an isotropic fifth-rank tensor. Show that the most general isotropic fifth-rank tensor must have at least ten independent components.

7.17 Show that the kinetic energy E of a body rotating with angular velocity $\boldsymbol{\Omega}$ is related to its inertia tensor I_{jk} by $E = I_{jk}\Omega_j\Omega_k/2$.

8
Applications of Vector Calculus

This chapter provides a brief introduction to some of the many applications of vector calculus to physics. Each of these is a vast topic in itself and is the subject of numerous books and a great deal of current research, so it is not possible to go into any detail in this book. However, a number of important governing equations and results can be obtained using the methods described in the previous chapters. In particular, it will be seen that the equations describing the behaviour of physical quantities such as electric fields and the velocity of a fluid are written in terms of the gradient, divergence and curl operators.

The following sections discuss the flow of heat within a body, the behaviour of electric and magnetic fields, the mechanics of solids and the mechanics of fluids. There are however several other subjects which use the language of vector calculus, including the theories of quantum mechanics and general relativity.

8.1 Heat transfer

In this section the equation describing the flow of heat within a solid body is derived. The argument is based on the law of conservation of energy, so is similar to the argument for the conservation of mass for a fluid given in Section 5.1.1.

Consider a solid with a temperature T which depends on space and time and a thermal conductivity K. Then the heat flows from hot to cold at a rate proportional to the temperature gradient, so the heat flux $\boldsymbol{q}$ is given by $\boldsymbol{q} = -K\boldsymbol{\nabla}T$. The minus sign appears here because the vector $\boldsymbol{\nabla}T$ points in the direction of increasing temperature but the heat flows in the direction of decreasing temperature.

Now consider an arbitrary region within the solid, denoted by a volume V with surface S and outward normal $\boldsymbol{n}$. The thermal energy or heat content of a volume element dV is $T\,c\,\rho\,dV$ where ρ is the density of the material and c is its specific heat. So the total heat content H of the volume V is

$$H = \iiint_V T\,c\,\rho\,dV.$$

The rate of change of this heat content must equal the rate at which heat flows into the volume V, assuming that there are no sources of heat within V. This rate of inflow of heat is the integral of the heat flux $-\boldsymbol{q}\cdot\boldsymbol{n}$ over the surface S, where the minus sign appears since for heat to flow in, $\boldsymbol{q}$ must point in the opposite direction to $\boldsymbol{n}$. Equating the rate of change of heat content with the rate at which heat flows into V gives

$$\iiint_V \frac{\partial T}{\partial t}\,c\,\rho\,dV = \oiint_S -\boldsymbol{q}\cdot\boldsymbol{n}\,dS = \oiint_S K\boldsymbol{\nabla}T\cdot\boldsymbol{n}\,dS.$$

The surface integral on the r.h.s. can be converted to a volume integral using the divergence theorem, giving

$$\iiint_V \frac{\partial T}{\partial t}\,c\,\rho\,dV = \iiint_V \boldsymbol{\nabla}\cdot(K\boldsymbol{\nabla}T)\,dV. \tag{8.1}$$

Finally, since the volume V is arbitrary, the volume integrals can be cancelled, giving

$$c\,\rho\frac{\partial T}{\partial t} = \boldsymbol{\nabla}\cdot(K\boldsymbol{\nabla}T), \tag{8.2}$$

since if (8.2) were not true at any point in space, then introducing a small volume V around this point would contradict (8.1).

If the thermal conductivity K is independent of position, then (8.2) can be simplified to give

$$\frac{\partial T}{\partial t} = k\,\nabla^2 T, \tag{8.3}$$

which is known as the *heat equation* or *diffusion equation*, and is of great importance since it occurs in many other contexts besides heat conduction. Other applications include processes of molecular diffusion such as the transport of chemicals within a living cell. The parameter $k = K/c\,\rho$ is known as the thermal diffusivity of the material. Note that the units of the diffusivity k are length^2/time.

If the system is steady, so that there is no dependence on time, then T obeys Laplace's equation, $\nabla^2 T = 0$.

The effect of the heat equation is to smooth out the distribution of temperature within a body. This is illustrated by the following examples.

Example 8.1

The surface S of a body is maintained at a constant temperature T_0. Show that the temperature T within the body approaches T_0 as $t \to \infty$.

Define a new temperature θ by $\theta = T - T_0$. Then since T_0 is fixed, θ obeys the heat equation

$$\frac{\partial \theta}{\partial t} = k\nabla^2\theta$$

within the volume V of the body and $\theta = 0$ on the boundary S. Multiplying through by θ gives

$$\frac{1}{2}\frac{\partial \theta^2}{\partial t} = k\theta\,\nabla^2\theta = k\boldsymbol{\nabla}\cdot(\theta\boldsymbol{\nabla}\theta) - k|\boldsymbol{\nabla}\theta|^2.$$

Now integrating over the volume V and applying the divergence theorem gives

$$\iiint_V \frac{1}{2}\frac{\partial \theta^2}{\partial t}\,dV = k\oint\!\!\!\oint_S \theta\boldsymbol{\nabla}\theta\cdot\boldsymbol{n}\,dS - k\iiint_V |\boldsymbol{\nabla}\theta|^2\,dV.$$

The surface integral is zero since $\theta = 0$ on S, so

$$\frac{1}{2}\frac{d}{dt}\iiint_V \theta^2\,dV = -k\iiint_V |\boldsymbol{\nabla}\theta|^2\,dV \le 0,$$

so the volume integral of θ^2 decreases unless θ is a constant. Since $\theta = 0$ on S the only possible value for this constant is zero, so the temperature decreases until $\theta = 0$, i.e. $T = T_0$ everywhere.

Example 8.2

Find a solution to the heat equation which is proportional to $\sin(ax)$, where a is a constant.

Seek a solution to (8.3) of the form $T(x,t) = f(t)\sin(ax)$. Substituting into (8.3) gives

$$\frac{df}{dt}\sin(ax) = -ka^2 f \sin(ax) \quad \Rightarrow \quad \frac{df}{dt} = -ka^2 f.$$

The solution to this differential equation for f is $f = f_0 \exp(-ka^2 t)$, where f_0 is a constant, so

$$T(x,t) = f_0 \exp(-ka^2 t)\sin(ax).$$

Note that the amplitude of the solution decreases exponentially with time, and that shorter waves (larger a) decay more rapidly than longer waves (smaller a).

8.2 Electromagnetism

The fundamental equations describing the behaviour of an electric field $\boldsymbol{E}$ and a magnetic field $\boldsymbol{B}$ are written in terms of the divergence and curl of these vector fields:

$$\nabla \cdot \boldsymbol{E} = \frac{\rho}{\epsilon_0}, \tag{8.4}$$

$$\nabla \cdot \boldsymbol{B} = 0, \tag{8.5}$$

$$\nabla \times \boldsymbol{E} = -\frac{\partial \boldsymbol{B}}{\partial t}, \tag{8.6}$$

$$\nabla \times \boldsymbol{B} = \mu_0 \boldsymbol{j} + \mu_0 \epsilon_0 \frac{\partial \boldsymbol{E}}{\partial t}, \tag{8.7}$$

where ρ is the density of electric charge, ϵ_0 and μ_0 are positive constants and $\boldsymbol{j}$ is the electric current density. These are known as Maxwell's equations and from these many important properties of electric and magnetic fields can be derived.

Each of the equations (8.4)–(8.7) can alternatively be written in an integral form. By integrating (8.4) over a volume V with surface S and applying the divergence theorem, we obtain

$$\oiint_S \boldsymbol{E} \cdot \boldsymbol{n}\, dS = \iiint_V \frac{\rho}{\epsilon_0}\, dV,$$

which is known as Gauss's law. In words, this states that the total flux of electric field out of the surface S is proportional to the amount of electric charge contained inside the surface. Similarly, from (8.5) it follows that

$$\oiint_S \boldsymbol{B} \cdot \boldsymbol{n}\, dS = 0,$$

so that the net flux of magnetic field through any closed surface is zero. By integrating (8.6) over a surface S and using Stokes's theorem, Faraday's law of electromagnetic induction is obtained:

$$\oint_C \boldsymbol{E} \cdot d\boldsymbol{r} = -\frac{\partial}{\partial t} \iint_S \boldsymbol{B} \cdot \boldsymbol{n}\, dS.$$

Finally, from (8.7) we obtain

$$\oint_C \boldsymbol{B} \cdot d\boldsymbol{r} = \mu_0 \iint_S \left(\boldsymbol{j} + \epsilon_0 \frac{\partial \boldsymbol{E}}{\partial t} \right) \cdot \boldsymbol{n}\, dS$$

which reduces to Ampère's law when there is no time dependence.

8.2.1 Electrostatics

Electrostatics is the study of steady electric fields. In a steady state, it follows from (8.6) that $\boldsymbol{E}$ is irrotational:

$$\boldsymbol{\nabla} \times \boldsymbol{E} = \boldsymbol{0}.$$

Recall from Example 5.5 and Theorem 3.1 that this means that $\boldsymbol{E}$ is conservative and that $\boldsymbol{E}$ can be written as the gradient of a potential, $\boldsymbol{E} = -\boldsymbol{\nabla}\Phi$. The choice of the minus sign here is merely a convention.

Now applying (8.4), the potential Φ must obey

$$\nabla^2 \Phi = -\boldsymbol{\nabla} \cdot \boldsymbol{E} = -\frac{\rho}{\epsilon_0}. \tag{8.8}$$

This equation is known as *Poisson's equation.* If the charge density ρ is zero, then Φ obeys Laplace's equation $\nabla^2\Phi = 0$. The problem of determining the electric field and potential due to a stationary distribution of electric charges reduces to the problem of solving Poisson's equation or Laplace's equation in the appropriate geometry. These problems are generally referred to as 'potential problems' and the associated theory is known as 'potential theory'. Some simple examples are given below.

Example 8.3

A sphere of radius a contains a uniform distribution of electric charge ρ. Determine the electric field $\boldsymbol{E}$ and the potential Φ both inside and outside the sphere.

Since the charge distribution is uniform within a sphere, it is most appropriate to use spherical polar coordinates and there is no preferred direction so both $\boldsymbol{E}$ and Φ only depend on the radial coordinate r. Referring back to (6.24) for the formula for the Laplacian in spherical polar coordinates, Poisson's equation for the potential inside the sphere is

$$\frac{1}{r^2}\frac{\partial}{\partial r}\left(r^2\frac{\partial \Phi}{\partial r}\right) = -\rho/\epsilon_0.$$

Multiplying by r^2 and integrating gives

$$r^2\frac{\partial \Phi}{\partial r} = -\rho r^3/3\epsilon_0 + C,$$

where C is a constant of integration, so

$$\frac{\partial \Phi}{\partial r} = -\rho r/3\epsilon_0 + C/r^2 = -E_r,$$

where E_r is the component of $\boldsymbol{E}$ in the r direction. Now since the electric field is a physical, measurable quantity, the constant C must be zero to avoid a singularity at the centre of the sphere $r = 0$. Within the sphere therefore the electric field and potential are:

$$E_r = \rho r/3\epsilon_0, \qquad \Phi = -\rho r^2/6\epsilon_0 + D,$$

where D is an arbitrary constant which always appears in potentials.

Outside the sphere, Laplace's equation holds,

$$\frac{1}{r^2}\frac{\partial}{\partial r}\left(r^2\frac{\partial \Phi}{\partial r}\right) = 0,$$

so

$$r^2\frac{\partial \Phi}{\partial r} = C \quad \Rightarrow \quad \frac{\partial \Phi}{\partial r} = C/r^2 = -E_r.$$

In this case the constant C cannot be set to zero since $r = 0$ is outside the region of consideration. Instead, C can be found by imposing that the electric field is continuous across the surface $r = a$. Equating the values of E_r at $r = a$ from the formulae for E_r inside and outside the sphere gives

$$\rho a/3\epsilon_0 = -C/a^2 \quad \Rightarrow \quad C = -\rho a^3/3\epsilon_0.$$

Hence, outside the sphere

$$E_r = \frac{\rho a^3}{3\epsilon_0 r^2}, \qquad \Phi = \frac{\rho a^3}{3\epsilon_0 r},$$

where the constant of integration associated with Φ has been set to zero. Note that the electric field obeys the inverse square law: E_r is proportional to r^{-2}.

Example 8.4

Find the electric field $\boldsymbol{E}$ and the potential Φ due to a point charge Q.

This can be done using the result of the previous example, by considering a point charge as a small sphere of uniform charge density. The total electric charge Q contained within a sphere of radius a with uniform charge density ρ is just the density multiplied by the volume, $Q = 4\pi a^3\rho/3$. The electric field and potential can then be written in terms of Q as

$$E_r = \frac{Q}{4\pi\epsilon_0 r^2}, \qquad \Phi = \frac{Q}{4\pi\epsilon_0 r}.$$

8.2.2 Electromagnetic waves in a vacuum

In a vacuum, where there is no electric charge and no electric current, Maxwell's equations take a simple and symmetric form:

$$\nabla \cdot \boldsymbol{E} = 0, \tag{8.9}$$

$$\nabla \cdot \boldsymbol{B} = 0, \tag{8.10}$$

$$\nabla \times \boldsymbol{E} = -\frac{\partial \boldsymbol{B}}{\partial t}, \tag{8.11}$$

$$\nabla \times \boldsymbol{B} = \mu_0\epsilon_0\frac{\partial \boldsymbol{E}}{\partial t}. \tag{8.12}$$

Taking the curl of (8.11) and using (4.23) gives

$$\nabla \times (\nabla \times \boldsymbol{E}) = \nabla(\nabla \cdot \boldsymbol{E}) - \nabla^2\boldsymbol{E} = -\nabla \times \frac{\partial \boldsymbol{B}}{\partial t}.$$

Now using (8.9) and (8.12) this can be written

$$-\nabla^2\boldsymbol{E} = -\frac{\partial}{\partial t}\left(\mu_0\epsilon_0\frac{\partial \boldsymbol{E}}{\partial t}\right),$$

so E obeys

$$\frac{\partial^2 \boldsymbol{E}}{\partial t^2} = c^2\nabla^2\boldsymbol{E} \tag{8.13}$$

where $c^2 = 1/\mu_0\epsilon_0$. This equation is called the ***wave equation*** since its solutions are waves travelling at speed c, as shown in the examples below. The constant c is the speed of light and the waves are known as electromagnetic waves.

Radio waves, light and X-rays are all examples of electromagnetic waves. These different types of waves have different frequencies but they all travel at the same speed c.

Example 8.5

Suppose that $\boldsymbol{E} = (f(x,t), 0, 0)$ in a Cartesian coordinate system so that the wave equation (8.13) becomes

$$\frac{\partial^2 f}{\partial t^2} = c^2 \frac{\partial^2 f}{\partial x^2}. \tag{8.14}$$

Show that $f(x,t) = \sin k(x - ct)$ is a solution for any value of the constant k and interpret this solution physically.

For the function $f(x,t) = \sin k(x - ct)$,

$$\frac{\partial^2 f}{\partial t^2} = -(kc)^2 \sin k(x - ct) \qquad \text{and} \qquad \frac{\partial^2 f}{\partial x^2} = -k^2 \sin k(x - ct),$$

so $f(x,t) = \sin k(x - ct)$ obeys (8.14). Physically, this solution corresponds to

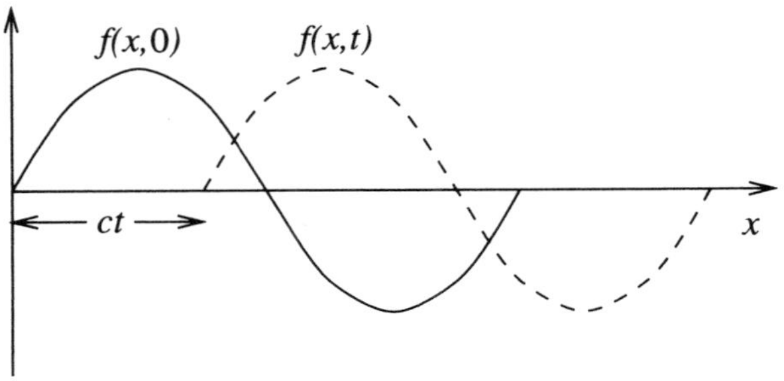

Fig. 8.1. The solution $f(x,t) = \sin k(x - ct)$ of the wave equation, at $t = 0$ (solid line) and at some later time (dashed line).

a sine wave travelling to the right at speed c. At $t = 0$, $f(x,0) = \sin kx$ which is a sine wave which has $f = 0$ at $x = 0$. At a later time t, the point at which $f = 0$ has moved to the position $x = ct$, so the wave has moved to the right a distance ct (Figure 8.1). The speed of the wave is therefore c. Similarly, it can be shown that the function $g(x,t) = \sin k(x + ct)$ is a solution of (8.14), representing a sine wave travelling to the left at speed c.

Example 8.6

Show that $\boldsymbol{E} = \boldsymbol{E}_0 f(\boldsymbol{k}\cdot\boldsymbol{x} - \omega t)$ is a solution to the wave equation (8.13), where $\boldsymbol{E}_0$ and $\boldsymbol{k}$ are constant vectors, f is any function and $\omega = \pm c|\boldsymbol{k}|$.

Let $u = \boldsymbol{k}\cdot\boldsymbol{x} - \omega t$ so that the solution under consideration is $\boldsymbol{E} = \boldsymbol{E}_0 f(u)$. Then

$$\frac{\partial \boldsymbol{E}}{\partial t} = \boldsymbol{E}_0 \frac{df}{du}\frac{\partial u}{\partial t} = -\omega \boldsymbol{E}_0 \frac{df}{du}$$

and similarly

$$\frac{\partial^2 \boldsymbol{E}}{\partial t^2} = -\omega \boldsymbol{E}_0 \frac{d^2 f}{du^2}\frac{\partial u}{\partial t} = \omega^2 \boldsymbol{E}_0 \frac{d^2 f}{du^2}.$$

To find $\nabla^2 \boldsymbol{E}$, note that since $\boldsymbol{E}_0$ is constant, $\nabla^2(\boldsymbol{E}_0 f(u)) = \boldsymbol{E}_0 \nabla^2 f(u) = \boldsymbol{E}_0 \boldsymbol{\nabla}\cdot(\boldsymbol{\nabla} f(u))$. The gradient of $f(u)$ is

$$\boldsymbol{\nabla} f(u) = \frac{df}{du}\boldsymbol{\nabla}(\boldsymbol{k}\cdot\boldsymbol{x} - \omega t) = \frac{df}{du}\boldsymbol{k},$$

and taking the divergence of this gives

$$\nabla^2 f(u) = \boldsymbol{\nabla}\left(\frac{df}{du}\right)\cdot\boldsymbol{k} = \frac{d^2 f}{du^2}\boldsymbol{\nabla} u\cdot\boldsymbol{k} = \frac{d^2 f}{du^2}|\boldsymbol{k}|^2.$$

Therefore, the wave equation is obeyed provided that

$$\omega^2 \boldsymbol{E}_0 \frac{d^2 f}{du^2} = c^2 \boldsymbol{E}_0 \frac{d^2 f}{du^2}|\boldsymbol{k}|^2 \quad \Rightarrow \quad \omega^2 = c^2|\boldsymbol{k}|^2.$$

So the function f is arbitrary and the only condition is that the frequency of the wave ω must be related to the speed c and the constant vector $\boldsymbol{k}$ (which is known as the wave vector) by $\omega = \pm c|\boldsymbol{k}|$.

EXERCISES

8.1 Use dimensional analysis to determine how the time taken for heat to diffuse through a body depends on the size L of the body and its thermal diffusivity k. Hence answer the following questions.
(a) If it takes six hours to defrost a frozen chicken, how long would it take to defrost a woolly mammoth?
(b) Cookery books state that the time taken to cook meat is, for example, twenty minutes per pound plus twenty minutes. Is this correct?

8.2 Show directly from Maxwell's equations that the charge density ρ and the electric current density $\boldsymbol{j}$ obey the conservation law

$$\frac{\partial \rho}{\partial t} + \boldsymbol{\nabla}\cdot\boldsymbol{j} = 0.$$

8.3 Derive the formula for the electric field $\boldsymbol{E}$ due to a point charge Q using Gauss's law.

8.4 Show from Maxwell's equations in a vacuum that the magnetic field $\boldsymbol{B}$ obeys the wave equation.

8.5 For the electromagnetic wave in which the electric field is given by $\boldsymbol{E} = \boldsymbol{E}_0 f(\boldsymbol{k} \cdot \boldsymbol{x} - \omega t)$, calculate the corresponding magnetic field $\boldsymbol{B}$. What can be deduced about the directions of the vectors $\boldsymbol{E}$ and $\boldsymbol{B}$?

8.6 Using Maxwell's equations in a vacuum, obtain an equation in the form of a conservation law for the rate of change of the energy $w = |\boldsymbol{B}|^2/2 + |\boldsymbol{E}|^2/2c^2$ of an electromagnetic wave.

8.3 Continuum mechanics and the stress tensor

Continuum mechanics is the study of continuous media including solids, liquids and gases. Solids have the property that when acted on by a force they deform but then reach an equilibrium in which the internal forces within the material balance the imposed force; this behaviour is described in Section 8.4. Fluids, which include liquids and gases, move continuously when subjected to a force and so require a different treatment (Section 8.5). Common to both solids and fluids is the concept of stress and the stress tensor described below.

Consider a small section of a surface, with area δS and unit normal $\boldsymbol{n}$, within a material (Figure 8.2). The two sides of the material are labelled side 1 and side 2, with $\boldsymbol{n}$ pointing in the direction of side 2. The material on side 2 exerts a force $\delta\boldsymbol{F}$ on the material on side 1 through the surface δS, due to interactions between the molecules of the material. Since the force is exerted through δS, the magnitude of the force is proportional to the area δS. The force is a vector quantity, and the only available vector is $\boldsymbol{n}$, so $\delta\boldsymbol{F}$ must be some quantity multiplied by $\boldsymbol{n}$. However, $\delta\boldsymbol{F}$ need not be parallel to $\boldsymbol{n}$. The angle between $\delta\boldsymbol{F}$ and $\boldsymbol{n}$ can be allowed to vary by using a tensor to relate $\delta\boldsymbol{F}$ and $\boldsymbol{n}$,

$$\delta F_i = P_{ij} n_j \, \delta S, \tag{8.15}$$

where P_{ij} is known as the *stress tensor* of the material. Since $\delta\boldsymbol{F}$ and $\boldsymbol{n}$ are vectors, it follows from the quotient rule that P_{ij} is a tensor.

Consider now the force exerted by side 1 on side 2. This is obtained by reversing the direction of the normal and using (8.15), so the force is $-P_{ij} n_j \, \delta S$.

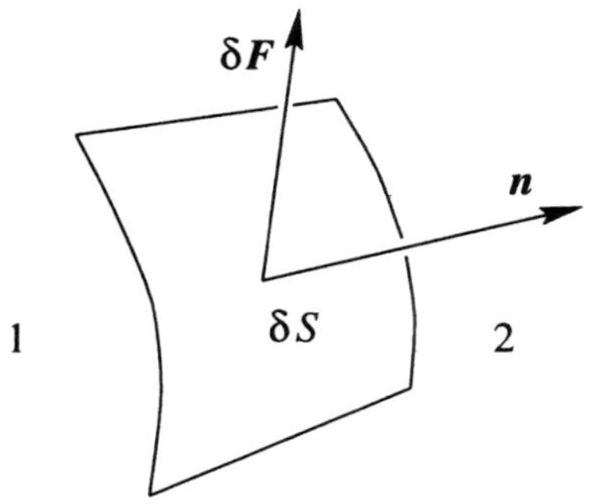

Fig. 8.2. A force $\delta \boldsymbol{F}$ is exerted on the material on side 1 through the surface δS.

This is just $-\delta F_i$, so this is consistent with Newton's third law as the forces are equal and opposite.

Now consider a region V within the material, with surface S and outward normal $\boldsymbol{n}$ (Figure 8.3). The total force exerted on this region by the surrounding

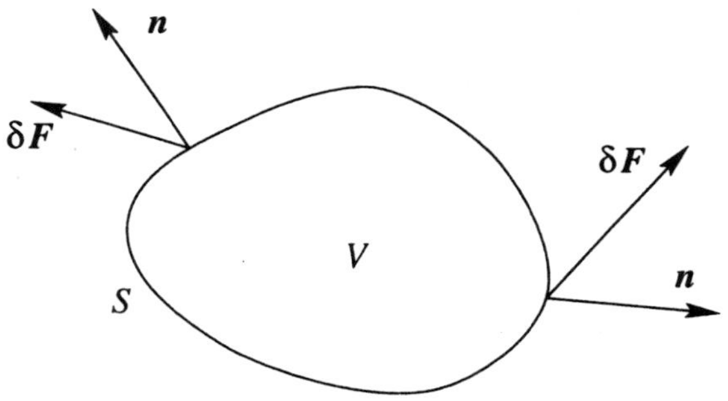

Fig. 8.3. The total force on a volume V is found by integrating over the surface S.

material is found by integrating the force due to all the surface elements over the surface S:

$$F_i = \oiint_S P_{ij} n_j \, dS = \iiint_V \frac{\partial P_{ij}}{\partial x_j} \, dV$$

using the divergence theorem for tensors (7.16). In general there may be other forces acting on the volume V, for example the force due to gravity; such long-range forces are known as body forces. Suppose, however, that body forces are negligible so that the only forces acting on the volume V are those exerted through the surface S. If the material is in equilibrium, then the total force acting on the region must be zero. Since this must be true for any arbitrary volume V, the stress tensor P_{ij} for a system in equilibrium must obey

$$\frac{\partial P_{ij}}{\partial x_j} = 0.$$

An additional constraint on P_{ij} comes from the consideration of the moment of the forces acting on a volume V. The moment of a force $\boldsymbol{F}$ about a point O is $\boldsymbol{r} \times \boldsymbol{F}$ where $\boldsymbol{r}$ is the position vector of the point of application of the force relative to O. The total moment of the forces acting through the surface S is therefore

$$\oiint_S \epsilon_{ijk} x_j \, dF_k = \oiint_S \epsilon_{ijk} x_j P_{km} n_m \, dS.$$

Using the tensor form of the divergence theorem again, this can be written as

$$\iiint_V \epsilon_{ijk} \frac{\partial}{\partial x_m} (x_j P_{km}) \, dV = \iiint_V \epsilon_{ijk} \left(\delta_{jm} P_{km} + x_j \frac{\partial P_{km}}{\partial x_m} \right) dV.$$

Now for a body in equilibrium we have already shown that the second term in the bracket is zero. The total moment of the forces must also be zero (otherwise the body would start to rotate), and since the volume V is arbitrary it follows that

$$\epsilon_{ijk} P_{kj} = 0.$$

This means that the tensor P_{kj} is symmetric, as shown in Example 7.8. Therefore the stress tensor for a material in equilibrium must obey the two constraints

$$\frac{\partial P_{ij}}{\partial x_j} = 0 \qquad \text{and} \qquad P_{ij} = P_{ji}. \tag{8.16}$$

Example 8.7

Suppose that a material is subjected to a body force $\boldsymbol{b}$ per unit volume. Find the total force $\boldsymbol{f}$ acting per unit volume.

For a volume V the total force is the volume integral of $\boldsymbol{f}$ over V. This is made up of the volume integral of the body force $\boldsymbol{b}$ plus the surface integral of the forces due to the stress tensor:

$$\iiint_V f_i \, dV = \iiint_V b_i \, dV + \oiint_S P_{ij} n_j \, dS = \iiint_V b_i + \frac{\partial P_{ij}}{\partial x_j} \, dV.$$

Since this is true for any volume V,

$$f_i = b_i + \frac{\partial P_{ij}}{\partial x_j}. \tag{8.17}$$

8.4 Solid mechanics

In the theory of the mechanics of a solid material it is generally assumed that there is a relationship between the stress tensor of the material and the strain of the material. The strain is the amount of stretching or deformation of the material. Suppose that, as a result of imposed forces, a material is deformed so that the material originally at position vector $\boldsymbol{r}$ is moved a distance $\boldsymbol{v}(\boldsymbol{r})$. Then the *strain tensor* E_{ij} is defined by

$$E_{ij} = \frac{1}{2}\left(\frac{\partial v_i}{\partial x_j} + \frac{\partial v_j}{\partial x_i}\right). \tag{8.18}$$

The simplest assumption is that there is a linear relation between the stress tensor and the strain tensor. This is known as Hooke's law. Since the stress tensor and the strain tensor are both second-rank tensors, the quantity relating them will in general be a fourth-rank tensor. If a material obeys the linear relationship

$$P_{ij} = c_{ijkl}E_{kl} \tag{8.19}$$

then it is said to be an ideal elastic solid.

If the further assumption is made that the material is isotropic, then the tensor c_{ijkl} is isotropic, so from Section 7.3.3,

$$\begin{aligned} P_{ij} &= (\lambda\delta_{ij}\delta_{kl} + \mu\delta_{ik}\delta_{jl} + \nu\delta_{il}\delta_{jk})E_{kl} \\ &= \lambda\delta_{ij}E_{kk} + \mu E_{ij} + \nu E_{ji} \\ &= \lambda\delta_{ij}E_{kk} + (\mu+\nu)E_{ij} \end{aligned} \tag{8.20}$$

since from its definition E_{ij} is symmetric. There are therefore only two independent constants in the relationship between stress and strain for an isotropic elastic solid. Since ν is arbitrary, we may take $\nu = \mu$ to obtain

$$P_{ij} = \lambda\delta_{ij}E_{kk} + 2\mu E_{ij}. \tag{8.21}$$

The two constants λ and μ are known as Lamé's constants. The inverse relationship, giving the strain in terms of the stress, can be found by setting $i = j$, giving

$$P_{ii} = 3\lambda E_{kk} + 2\mu E_{ii} = (3\lambda + 2\mu)E_{kk}$$

and then substituting $E_{kk} = P_{kk}/(3\lambda + 2\mu)$ in (8.21), giving

$$E_{ij} = \frac{1}{2\mu}P_{ij} - \frac{\lambda\delta_{ij}P_{kk}}{(2\mu)(3\lambda + 2\mu)}.$$

Example 8.8

Find the strain tensor and the stress tensor for an isotropic elastic material when it is subjected to

(a) a stretching deformation $\boldsymbol{v} = (0, 0, ax_3)$;

(b) a shearing deformation $\boldsymbol{v} = (bx_3, 0, 0)$.

The two deformations are illustrated in Figure 8.4.

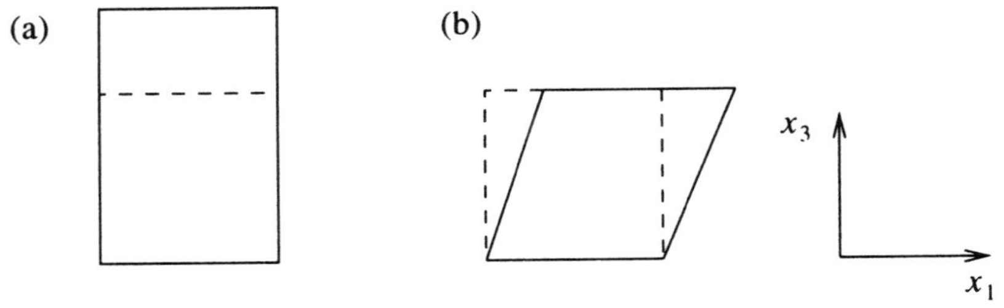

Fig. 8.4. Deformation of a cube of material (dashed line) by (a) $\boldsymbol{v} = (0, 0, ax_3)$, (b) $\boldsymbol{v} = (bx_3, 0, 0)$.

(a) Here the only non-zero element of E_{ij} is $E_{33} = a$. Hence $E_{kk} = a$ and so the non-zero elements of P_{ij} are $P_{11} = P_{22} = \lambda a$ and $P_{33} = (\lambda + 2\mu)a$.

(b) In this case $E_{31} = E_{13} = b/2$ and other elements of E_{ij} are zero. Since $E_{kk} = 0$ the only non-zero elements of P_{ij} are $P_{31} = P_{13} = \mu b$.

These two simple examples show that the diagonal components of P_{ij} represent forces of stretching or compression of the material, while the off-diagonal components represent shearing forces.

Example 8.9

Show that the anti-symmetric tensor

$$S_{ij} = \frac{1}{2}\left(\frac{\partial v_i}{\partial x_j} - \frac{\partial v_j}{\partial x_i}\right)$$

represents a rotation of the material.

Since S_{ij} is anti-symmetric, $S_{11} = S_{22} = S_{33} = 0$ and S_{ij} has only three independent components, S_{12}, S_{23} and S_{31}. The first of these is

$$S_{12} = \frac{1}{2}\left(\frac{\partial v_1}{\partial x_2} - \frac{\partial v_2}{\partial x_1}\right) = -\frac{1}{2}\left[\boldsymbol{\nabla} \times \boldsymbol{v}\right]_3 = -\frac{1}{2}\epsilon_{12k}\left[\boldsymbol{\nabla} \times \boldsymbol{v}\right]_k .$$

Similar equations hold for the other components, so S_{ij} is related to $\boldsymbol{\nabla} \times \boldsymbol{v}$ by

$$S_{ij} = -\frac{1}{2}\epsilon_{ijk}\left[\nabla \times \boldsymbol{v}\right]_k . \tag{8.22}$$

Thus the components of S_{ij} depend only on $\nabla \times \boldsymbol{v}$, which from Section 3.4.2 represents a rotation of the body.

Example 8.10

Show that in an isotropic material P_{ij} cannot depend on the anti-symmetric tensor S_{ij}.

Following the arguments above, if $P_{ij} = c_{ijkl}S_{kl}$ and the material is isotropic, then the analogue of (8.20) is

$$P_{ij} = \lambda\delta_{ij}S_{kk} + (\mu - \nu)S_{ij}.$$

Since S_{ij} is anti-symmetric, $S_{kk} = 0$. This means that the symmetric tensor P_{ij} is equal to a constant multiplied by the anti-symmetric tensor S_{ij}, which is clearly a contradiction.

8.5 Fluid mechanics

An important property which distinguishes a fluid (such as a liquid or a gas) from a solid is that a fluid is unable to support a shear stress. This means that if a fluid is subject to a shear stress, such as in Example 8.8(b), the fluid will move continuously. For a fluid at rest, therefore, the stress tensor can only include diagonal elements. Furthermore, if the fluid is subjected to a stretching force, as in Example 8.8(a), the fluid will stretch and continue to move for as long as the force acts. This motion can only be avoided if the forces acting on the fluid are the same in all directions. Hence the stress tensor in a fluid at rest must be isotropic,

$$P_{ij} = -p\,\delta_{ij}, \tag{8.23}$$

where p is the hydrostatic pressure of the fluid, which in general is a function of position. The pressure can then be defined in terms of the stress tensor by

$$p = -P_{jj}/3. \tag{8.24}$$

Now suppose that the fluid is in motion, with a velocity given by the vector field $\boldsymbol{u}(\boldsymbol{r})$. In this case, there is an additional contribution to the stress tensor due to the motion, so

$$P_{ij} = -p\,\delta_{ij} + d_{ij} \tag{8.25}$$

where $d_{ij} = 0$ when the fluid is at rest. Since (8.25) does not uniquely define p and d_{ij}, it can be imposed that (8.24) holds for a fluid in motion, which means that $d_{jj} = 0$.

The tensor d_{ij} represents the forces due to the fluid motion. These forces arise due to friction between adjacent molecules of the fluid that are moving at different velocities. It is generally assumed that there is a linear relationship between d_{ij} and the tensor representing the difference in velocities, $\partial u_i/\partial x_j$. A fluid behaving in this way is said to be Newtonian, and experiments show that this is a good approximation for most fluids. The analysis of the relationship between d_{ij} and $\partial u_i/\partial x_j$ now follows exactly as for the case of a solid (Section 8.4). If it is assumed that the fluid is isotropic, then d_{ij} can only depend on the symmetric part of $\partial u_i/\partial x_j$,

$$e_{ij} = \frac{1}{2}\left(\frac{\partial u_i}{\partial x_j} + \frac{\partial u_j}{\partial x_i}\right),$$

which is known as the rate-of-strain tensor. The argument of Example 8.10 shows that d_{ij} cannot depend on the anti-symmetric part of $\partial u_i/\partial x_j$. The form of the dependence of d_{ij} on e_{ij} is

$$d_{ij} = \lambda\delta_{ij}e_{kk} + 2\mu e_{ij}, \tag{8.26}$$

where, as in (8.21), λ and μ are constants. The constraint $d_{jj} = 0$ means that $3\lambda + 2\mu = 0$. Eliminating λ, the form of the stress tensor for a Newtonian fluid is then

$$P_{ij} = -p\,\delta_{ij} + 2\mu e_{ij} - \frac{2}{3}\mu\delta_{ij}e_{kk}. \tag{8.27}$$

The constant μ is the *viscosity* of the fluid, which represents the fluid's 'stickiness' or friction.

8.5.1 Equation of motion for a fluid

Having obtained the formula for the stress tensor in a fluid, it is now possible to write down the equation of motion. The equation of motion is derived from Newton's second law, force = mass × acceleration. The force can be obtained from the stress tensor using (8.17). The force $\boldsymbol{f}$ per unit volume due to body forces $\boldsymbol{b}$ and the stress tensor P_{ij} is

$$\begin{aligned}
f_i &= b_i + \frac{\partial P_{ij}}{\partial x_j} \\
&= b_i - \frac{\partial p}{\partial x_i} + \mu\left(\frac{\partial^2 u_i}{\partial x_j \partial x_j} + \frac{\partial^2 u_j}{\partial x_i \partial x_j}\right) - \frac{2}{3}\mu\frac{\partial^2 u_j}{\partial x_i \partial x_j} \\
&= b_i - \frac{\partial p}{\partial x_i} + \mu\frac{\partial^2 u_i}{\partial x_j \partial x_j} + \frac{1}{3}\mu\frac{\partial^2 u_j}{\partial x_i \partial x_j}.
\end{aligned}$$

In vector notation this can be written

$$\boldsymbol{f} = \boldsymbol{b} - \boldsymbol{\nabla} p + \mu \nabla^2 \boldsymbol{u} + \frac{1}{3}\mu \boldsymbol{\nabla}\boldsymbol{\nabla} \cdot \boldsymbol{u}.$$

To obtain the acceleration of a fluid particle, consider the change in velocity of a particle in a short time interval δt. A particle at position $\boldsymbol{r}$ moving with velocity $\boldsymbol{u}(\boldsymbol{r}, t)$ moves a distance $\boldsymbol{u}(\boldsymbol{r}, t)\delta t$ in this time. Therefore its new velocity is $\boldsymbol{u}(\boldsymbol{r} + \boldsymbol{u}(\boldsymbol{r}, t)\delta t, t + \delta t)$. The velocity is a function of four variables, $\boldsymbol{u} = \boldsymbol{u}(x, y, z, t)$, so expanding this new velocity using Taylor's theorem gives

$$\begin{aligned} \boldsymbol{u}(\boldsymbol{r} + \boldsymbol{u}(\boldsymbol{r}, t)\delta t, t + \delta t) &= \boldsymbol{u}(\boldsymbol{r}, t) + u_x \delta t \frac{\partial \boldsymbol{u}}{\partial x} + u_y \delta t \frac{\partial \boldsymbol{u}}{\partial y} + u_z \delta t \frac{\partial \boldsymbol{u}}{\partial z} + \delta t \frac{\partial \boldsymbol{u}}{\partial t} \\ &= \boldsymbol{u}(\boldsymbol{r}, t) + \delta t\, \boldsymbol{u} \cdot \boldsymbol{\nabla} \boldsymbol{u} + \delta t \frac{\partial \boldsymbol{u}}{\partial t}. \end{aligned}$$

The acceleration $\boldsymbol{a}$ of the particle is the limit of the difference between the new and old velocities divided by δt, so

$$\boldsymbol{a} = \frac{\partial \boldsymbol{u}}{\partial t} + \boldsymbol{u} \cdot \boldsymbol{\nabla} \boldsymbol{u}.$$

Since the mass per unit volume of the fluid is just the density ρ, the equation of motion is

$$\rho \left(\frac{\partial \boldsymbol{u}}{\partial t} + \boldsymbol{u} \cdot \boldsymbol{\nabla} \boldsymbol{u} \right) = \boldsymbol{b} - \boldsymbol{\nabla} p + \mu \nabla^2 \boldsymbol{u} + \frac{1}{3}\mu \boldsymbol{\nabla}\boldsymbol{\nabla} \cdot \boldsymbol{u}. \tag{8.28}$$

This important equation is known as the *Navier–Stokes equation.*

From now on, assume that the fluid is of constant density, so that ρ does not depend on time or space. Then, from Section 5.1.1, where the equation for conservation of mass for a fluid was derived, the velocity field $\boldsymbol{u}$ obeys $\boldsymbol{\nabla} \cdot \boldsymbol{u} = 0$. There are two more important equations of fluid mechanics that can be derived from the Navier–Stokes equation; these are described in the following sections.

8.5.2 The vorticity equation

Assume that the body force $\boldsymbol{b}$ is conservative, so that it may be written as the gradient of a potential, $\boldsymbol{b} = -\boldsymbol{\nabla}\Phi$. By using the vector identity

$$\boldsymbol{u} \cdot \boldsymbol{\nabla} \boldsymbol{u} = \boldsymbol{\nabla}(|\mathbf{u}|^2/2) - \boldsymbol{u} \times \boldsymbol{\nabla} \times \boldsymbol{u}$$

which was derived in Example 4.14, the Navier–Stokes equation can be written

$$\rho \left(\frac{\partial \boldsymbol{u}}{\partial t} + \boldsymbol{\nabla}(|\mathbf{u}|^2/2) - \boldsymbol{u} \times \boldsymbol{\omega} \right) = -\boldsymbol{\nabla} p - \boldsymbol{\nabla}\Phi + \mu \nabla^2 \boldsymbol{u}, \tag{8.29}$$

where $\boldsymbol{\omega} = \boldsymbol{\nabla} \times \boldsymbol{u}$ is the ***vorticity*** of the fluid, which as shown in Section 3.4.2 is proportional to the local rate of rotation of the fluid. Now by taking the curl of (8.29), the equation can be simplified considerably because the three terms involving grad will then disappear. The pressure and the body forces are then eliminated and an equation for the rate of change of the vorticity is obtained:

$$\frac{\partial \boldsymbol{\omega}}{\partial t} - \boldsymbol{\nabla} \times (\boldsymbol{u} \times \boldsymbol{\omega}) = \frac{\mu}{\rho}\nabla^2 \boldsymbol{\omega}, \tag{8.30}$$

where we have made use of the fact that the order of the curl and Laplacian operators may be interchanged. By expanding the curl of the cross product using (4.30) and noting that both $\boldsymbol{u}$ and $\boldsymbol{\omega}$ are solenoidal, (8.30) can be alternatively written

$$\frac{\partial \boldsymbol{\omega}}{\partial t} + \boldsymbol{u} \cdot \boldsymbol{\nabla} \boldsymbol{\omega} - \boldsymbol{\omega} \cdot \boldsymbol{\nabla} \boldsymbol{u} = \frac{\mu}{\rho}\nabla^2 \boldsymbol{\omega}. \tag{8.31}$$

This is known as the ***vorticity equation.***

Example 8.11

Obtain the simplified form of the vorticity equation for two-dimensional flow, $\boldsymbol{u}(\boldsymbol{r}) = (u(x, y), v(x, y), 0)$, of an incompressible fluid.

If $\boldsymbol{u} = (u(x, y), v(x, y), 0)$, then the vorticity $\boldsymbol{\omega}$ is

$$\boldsymbol{\omega} = \boldsymbol{\nabla} \times \boldsymbol{u} = (0, 0, \frac{\partial v}{\partial x} - \frac{\partial u}{\partial y}) = (0, 0, \omega(x, y)).$$

Thus the vorticity vector has only one component, which is perpendicular to the plane of motion of the fluid. This means that the term $\boldsymbol{\omega} \cdot \boldsymbol{\nabla} \boldsymbol{u}$ in the vorticity equation is zero, since $\boldsymbol{\omega} \cdot \boldsymbol{\nabla} \boldsymbol{u} = \omega \partial \boldsymbol{u}/\partial z = \boldsymbol{0}$. The vorticity equation simplifies to the scalar equation

$$\frac{\partial \omega}{\partial t} + \boldsymbol{u} \cdot \boldsymbol{\nabla} \omega = \frac{\mu}{\rho}\nabla^2 \omega. \tag{8.32}$$

Example 8.12

An incompressible fluid flows along a straight two-dimensional channel and the velocity $\boldsymbol{u}$ is parallel to the walls of the channel. Show that the vorticity equation reduces to the diffusion equation.

Choosing the x-axis to be parallel to the channel walls, the velocity $\boldsymbol{u}$ has the form $\boldsymbol{u} = (u, 0, 0)$. As the fluid is incompressible, $\boldsymbol{\nabla} \cdot \boldsymbol{u} = 0$ so $\partial u/\partial x = 0$. Since u is independent of x, so is ω, so the second term in (8.32) is $\boldsymbol{u} \cdot \boldsymbol{\nabla} \omega = u \partial \omega/\partial x = 0$. The vorticity equation is then

$$\frac{\partial \omega}{\partial t} = \frac{\mu}{\rho}\nabla^2 \omega,$$

which is the same as the diffusion equation (8.3).

8.5.3 Bernoulli's equation

A second important equation can be obtained by taking the dot product of $\boldsymbol{u}$ with (8.29). If the flow is steady, so the time derivative is zero, and the viscosity of the fluid is negligible, then

$$\boldsymbol{u} \cdot \boldsymbol{\nabla}(\rho|\boldsymbol{u}|^2/2 + p + \Phi) = 0. \tag{8.33}$$

This means that the quantity $\rho|\boldsymbol{u}|^2/2 + p + \Phi$ does not change in the direction of $\boldsymbol{u}$, so that it is constant along the path of a fluid particle. If in addition, the vorticity is zero, then (8.29) becomes

$$\boldsymbol{\nabla}(\rho|\boldsymbol{u}|^2/2 + p + \Phi) = \boldsymbol{0} \tag{8.34}$$

so $\rho|\boldsymbol{u}|^2/2 + p + \Phi$ is constant everywhere. These two results are different forms of *Bernoulli's equation.* In words, Bernoulli's equation means that where the velocity of a fluid is high, its pressure is low. This important result can be used to explain a wide range of physical observations, including the swerving of a spinning ball and the lift generated by an aeroplane's wing.

Example 8.13

Why does a spinning ball swerve?

Consider a ball travelling to the left with velocity $\boldsymbol{u}$. Equivalently, the ball may be considered to be stationary, with air flowing past it with velocity $\boldsymbol{v} = -\boldsymbol{u}$. Now suppose that the ball is spinning anticlockwise, as shown in Figure 8.5. As the ball spins, it drags air with it, rotating around the ball. The two components of the fluid motion due to the translation and the rotation are oppositely directed above the ball but point in the same direction below the ball. Therefore the total speed of the fluid is lower above the ball than below the ball. According to Bernoulli's equation, the pressure is greater above the ball than below. This pressure difference exerts a force on the ball which is directed downwards in Figure 8.5, causing the ball to swerve.

Example 8.14

To model the effect described qualitatively in the previous example, consider a cylinder of radius b and assume that in the frame of the cylinder the velocity at the surface of the cylinder is $(v \sin\theta - a)\boldsymbol{e}_\theta$ in polar coordinates. Assuming that the viscosity is negligible and that Bernoulli's equation is valid, find the pressure at the surface and hence calculate the total force acting on the cylinder.

If Bernoulli's equation holds and there are no body forces, then $p = c - \rho|\boldsymbol{u}|^2/2$ where c is constant, so

$$p = c - \rho(v^2 \sin^2\theta - 2av\sin\theta + a^2)/2.$$

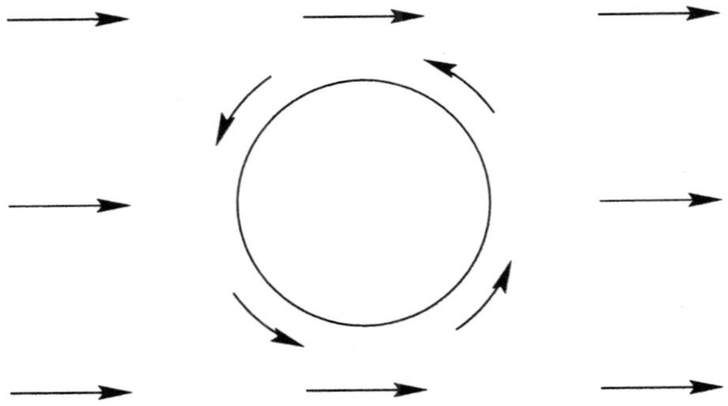

Fig. 8.5. A ball moving to the left and spinning anticlockwise experiences a net downward force.

Now if there is no viscosity the stress tensor is simply $P_{ij} = -p\,\delta_{ij}$ so the total force is

$$F_i = \oiint_S P_{ij} n_j \, dS = \oiint_S -p\, n_i \, dS.$$

In a Cartesian coordinate system (x_1, x_2), $\boldsymbol{n} = (\cos\theta, \sin\theta)$ and the area element dS is $b\,d\theta$ per unit length of the cylinder. The force per unit length in the x_2 direction is

$$\begin{aligned} F_2 &= \int_0^{2\pi} -(c - \rho(v^2 \sin^2\theta - 2av\sin\theta + a^2)/2)\sin\theta\, b\, d\theta \\ &= \int_0^{2\pi} -\rho a v \sin^2\theta\, b\, d\theta \\ &= -\pi\rho a v b. \end{aligned}$$

The force in the x_1 direction is zero since the functions $\sin^2\theta\cos\theta$, $\sin\theta\cos\theta$ and $\cos\theta$ all integrate to zero. This means that there is no drag force on the cylinder, which shows that this simple model is inadequate in some way. In fact it is the neglect of viscosity which is not valid. Even though the viscosity of a fluid such as air is small, it can have a large effect on the solution.

Summary of Chapter 8

- Many important equations of mathematical physics are written in terms of grad, div and curl.
- The heat equation, or diffusion equation, is

$$\frac{\partial T}{\partial t} = k\,\nabla^2 T.$$

This describes heat transfer within a body and other physical processes involving molecular diffusion.
- Electric and magnetic fields obey Maxwell's equations (8.4)–(8.7). In a steady state the electric field is irrotational and its potential Φ obeys Poisson's equation $\nabla^2\Phi = -\rho/\epsilon_0$ where ρ is the charge density and ϵ_0 is a constant. In a vacuum, Maxwell's equations lead to the wave equation which describes the motion of electromagnetic waves.
- In a continuous material such as a solid, a liquid or a gas, molecular forces are transferred through a surface δS with normal $\boldsymbol{n}$ according to the formula $\delta F_i = P_{ij} n_j\,\delta S$, where P_{ij} is the stress tensor of the material. If the material is in equilibrium, P_{ij} must obey

$$\frac{\partial P_{ij}}{\partial x_j} = 0 \qquad \text{and} \qquad P_{ij} = P_{ji}.$$

- In an isotropic, ideal elastic solid with a deformation $\boldsymbol{v}(\boldsymbol{r})$ the strain tensor E_{ij} and stress tensor P_{ij} are given by

$$E_{ij} = \frac{1}{2}\left(\frac{\partial v_i}{\partial x_j} + \frac{\partial v_j}{\partial x_i}\right), \qquad P_{ij} = \lambda\delta_{ij}E_{kk} + 2\mu E_{ij},$$

where λ and μ are constants.
- In a Newtonian fluid moving with velocity $\boldsymbol{u}(\boldsymbol{r})$, the rate-of-strain tensor e_{ij} and stress tensor P_{ij} are

$$e_{ij} = \frac{1}{2}\left(\frac{\partial u_i}{\partial x_j} + \frac{\partial u_j}{\partial x_i}\right), \qquad P_{ij} = -p\,\delta_{ij} + 2\mu e_{ij} - \frac{2}{3}\mu\delta_{ij}e_{kk},$$

where p is the pressure in the fluid and μ is its viscosity.
- If the fluid is incompressible then $\boldsymbol{\nabla}\cdot\boldsymbol{u} = 0$ and its equation of motion is the Navier–Stokes equation

$$\rho\left(\frac{\partial \boldsymbol{u}}{\partial t} + \boldsymbol{u}\cdot\boldsymbol{\nabla}\boldsymbol{u}\right) = \boldsymbol{b} - \boldsymbol{\nabla}p + \mu\nabla^2\boldsymbol{u},$$

where ρ is the density and $\boldsymbol{b}$ is the body force acting.

EXERCISES

8.7 A material has a stress tensor P_{ij} that takes the form

$$P_{ij} = \begin{pmatrix} 0 & a & x_2 \\ 0 & x_2^m & bx_2x_3 \\ x_2 & cx_2x_3 & x_3^n \end{pmatrix}$$

in a Cartesian coordinate system (x_1, x_2, x_3).

(a) If there are no body forces acting and the material is in equilibrium, find the values of the constants a, b, c, m and n.

(b) For these values of the constants, find the magnitude and direction of the force exerted on the surface $x_1 = 1$, $0 \le x_2, x_3 \le 1$, by the material in the region $x_1 > 1$.

8.8 Verify (8.22) by expanding the r.h.s. using suffix notation.

8.9 Show that for an isotropic elastic solid in equilibrium, the deformation $\boldsymbol{v}$ must obey $(\lambda + \mu)\boldsymbol{\nabla}\boldsymbol{\nabla} \cdot \boldsymbol{v} + \mu\nabla^2\boldsymbol{v} = 0$.

8.10 An isotropic elastic solid with Lamé constants λ and μ is subjected to a deformation $v_1 = ax_1x_2$, $v_2 = b(x_1^2 - x_2^2)$, $v_3 = 0$.

(a) Find the strain tensor E_{ij}.

(b) Find the stress tensor P_{ij}.

(c) Determine whether it is possible for the material to be in equilibrium.

8.11 A compressible fluid with negligible viscosity is initially at rest with uniform density ρ_0 and pressure p_0, with no body forces. A small perturbation is then introduced so that there is a velocity $\boldsymbol{u}(\boldsymbol{r}, t)$ and the density becomes $\rho_0 + \rho_1(\boldsymbol{r}, t)$.

(a) Assuming that products of the small quantities $\boldsymbol{u}$ and ρ_1 can be neglected, show that the equation for conservation of mass (5.9) becomes

$$\frac{\partial \rho_1}{\partial t} + \rho_0 \boldsymbol{\nabla} \cdot \boldsymbol{u} = 0. \tag{8.35}$$

(b) Assuming that the perturbation p_1 to the pressure is related to ρ_1 by $p_1 = a\rho_1$ where a is constant, show that the Navier–Stokes equation reduces to

$$\rho_0 \frac{\partial \boldsymbol{u}}{\partial t} = -a\boldsymbol{\nabla}\rho_1. \tag{8.36}$$

(c) Hence show that the density perturbation ρ_1 obeys the wave equation and interpret this result physically.

Solutions

Solutions to Exercises for Chapter 1

1.1 Density and power are scalars; all the others are vectors.

1.2 $|\boldsymbol{a}| = \sqrt{4+0+9} = \sqrt{13}$, $|\boldsymbol{b}| = \sqrt{1+0+1} = \sqrt{2}$, $\boldsymbol{a}+\boldsymbol{b} = (3,0,2)$, $\boldsymbol{a}-\boldsymbol{b} = (1,0,4)$ and $\boldsymbol{a}\cdot\boldsymbol{b} = 2+0-3 = -1$. To find the angle between $\boldsymbol{a}$ and $\boldsymbol{b}$, use the formula $\boldsymbol{a}\cdot\boldsymbol{b} = |\boldsymbol{a}||\boldsymbol{b}|\cos\theta$. Substituting the values already obtained, this becomes $-1 = \sqrt{13}\sqrt{2}\cos\theta$, so $\cos\theta \approx -0.196$ and $\theta \approx 101°$.

1.3 The component of $\boldsymbol{u}$ in the direction of $\boldsymbol{v}$ is $\boldsymbol{u}\cdot\boldsymbol{v}/|\boldsymbol{v}|$. For $\boldsymbol{u} = (1,2,2)$ and $\boldsymbol{v} = (-6,2,3)$, $\boldsymbol{u}\cdot\boldsymbol{v} = -6+4+6 = 4$ and $|\boldsymbol{v}| = \sqrt{36+4+9} = 7$, so the answer is $4/7$. Similarly the component of $\boldsymbol{v}$ in the direction of $\boldsymbol{u}$ is $\boldsymbol{u}\cdot\boldsymbol{v}/|\boldsymbol{u}| = 4/3$.

1.4 The plane perpendicular to $\boldsymbol{a} = (1,1,-1)$ is $\boldsymbol{r}\cdot\boldsymbol{a} =$ constant, so $x+y-z =$ constant. If the plane passes through $x = 1$, $y = 2$, $z = 1$ then the value of the constant is $1+2-1 = 2$.

1.5 Let two adjacent sides of the rhombus be the vectors $\boldsymbol{a}$ and $\boldsymbol{b}$ (as in Figure 1.10). Then since the sides are of equal length, $|\boldsymbol{a}| = |\boldsymbol{b}|$. The diagonals of the rhombus are $\boldsymbol{a}+\boldsymbol{b}$ and $\boldsymbol{a}-\boldsymbol{b}$. Taking the dot product of the two diagonals, $(\boldsymbol{a}+\boldsymbol{b})\cdot(\boldsymbol{a}-\boldsymbol{b}) = \boldsymbol{a}\cdot\boldsymbol{a}+\boldsymbol{b}\cdot\boldsymbol{a}-\boldsymbol{a}\cdot\boldsymbol{b}-\boldsymbol{b}\cdot\boldsymbol{b} = |\boldsymbol{a}|^2-|\boldsymbol{b}|^2 = 0$, so the diagonals are perpendicular.

1.6 Consider the unit cube, which has sides of length 1 parallel to the coordinate axes. Two opposite vertices are the points $(0,0,0)$ and $(1,1,1)$, so one diagonal is the vector $(1,1,1)$. Another diagonal runs from $(0,0,1)$ to $(1,1,0)$, so another diagonal is the vector $(1,1,-1)$. Both diagonals have magnitude

$\sqrt{3}$ and their dot product is 1. Using the formula $\boldsymbol{a} \cdot \boldsymbol{b} = |\boldsymbol{a}||\boldsymbol{b}| \cos\theta$ gives $1 = 3\cos\theta$, so $\theta = \cos^{-1}(1/3) \approx 70.5°$.

1.7 Choose the origin at one vertex of the triangle and let two of the sides be represented by the vectors $\boldsymbol{a}$ and $\boldsymbol{b}$. Then the third side is represented by the vector $\boldsymbol{b} - \boldsymbol{a}$. The midpoint of the side opposite the origin has the position

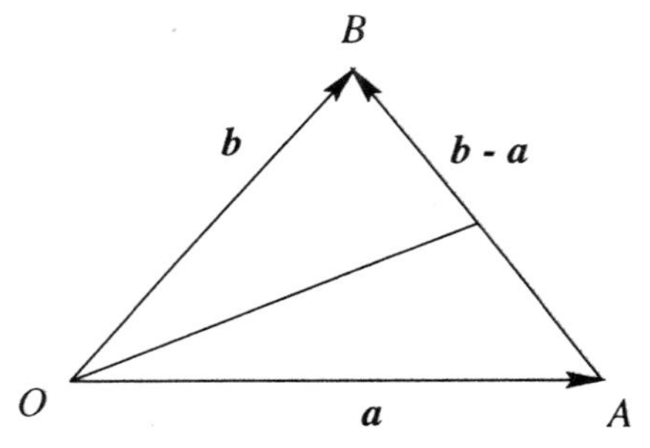

vector $\boldsymbol{a} + (\boldsymbol{b} - \boldsymbol{a})/2 = (\boldsymbol{a} + \boldsymbol{b})/2$, so the line from the origin to the midpoint of the opposite side is given parametrically by $\boldsymbol{r} = \lambda(\boldsymbol{a} + \boldsymbol{b})/2$. Similarly the line from A to the midpoint of the opposite side is given parametrically by $\boldsymbol{r} = \boldsymbol{a} + \mu(\boldsymbol{b}/2 - \boldsymbol{a})$. These two lines meet when $\lambda(\boldsymbol{a} + \boldsymbol{b})/2 = \boldsymbol{a} + \mu(\boldsymbol{b}/2 - \boldsymbol{a})$. In this equation we can equate the terms in $\boldsymbol{a}$ and $\boldsymbol{b}$. This gives the solution $\mu = \lambda = 2/3$ so the lines meet at the point with position vector $(\boldsymbol{a} + \boldsymbol{b})/3$. Similarly, it can be found that the line from B to the midpoint of the opposite side meets the line from the origin to the midpoint of the opposite side at $(\boldsymbol{a} + \boldsymbol{b})/3$, so all three lines pass through this point.

1.8 The position vector $(1,1,1)$ lies on the line and a vector along the line is $(2,3,5) - (1,1,1) = (1,2,4)$. The parametric form is therefore $\boldsymbol{r} = (1,1,1) + \lambda(1,2,4)$. The cross product form is found by taking the cross product of this equation with $(1,2,4)$: $\boldsymbol{r} \times (1,2,4) = (1,1,1) \times (1,2,4) = (2,-3,1)$.

1.9 To prove the sine rule, use the fact that the area of the triangle is $|\boldsymbol{a} \times \boldsymbol{b}|/2$. Since this area is also equal to $|\boldsymbol{b} \times \mathbf{c}|/2$ and $|\boldsymbol{c} \times \boldsymbol{a}|/2$,

$$ab \sin C = bc \sin A = ca \sin B,$$

and dividing by abc gives the sine rule.
The cosine rule follows from the magnitude of the vector $\mathbf{c}$:

$$c^2 = |\mathbf{c}|^2 = \mathbf{c} \cdot \mathbf{c} = (\mathbf{a} - \mathbf{b}) \cdot (\mathbf{a} - \mathbf{b}) = a^2 + b^2 - 2\mathbf{a} \cdot \mathbf{b} = a^2 + b^2 - 2ab \cos C.$$

1.10 (a) (i) For any two vectors $\boldsymbol{a}$ and $\boldsymbol{b}$, $\boldsymbol{a} + \boldsymbol{b}$ is a vector; (ii) $\boldsymbol{a} + (\boldsymbol{b} + \boldsymbol{c}) = (\boldsymbol{a} + \boldsymbol{b}) + \boldsymbol{c}$; (iii) $\boldsymbol{a} + \mathbf{0} = \mathbf{0} + \boldsymbol{a} = \boldsymbol{a}$; (iv) $\boldsymbol{a} + (-\boldsymbol{a}) = (-\boldsymbol{a}) + \boldsymbol{a} = \mathbf{0}$. Hence vectors and addition form a group.

(b) Vectors and the dot product do not form a group since $\boldsymbol{a}\cdot\boldsymbol{b}$ is not a vector.

(c) Vectors and the cross product do not form a group because $\boldsymbol{a}\times(\boldsymbol{b}\times\boldsymbol{c})\neq(\boldsymbol{a}\times\boldsymbol{b})\times\boldsymbol{c}$. Also, there is no identity element.

1.11 (a) $|\boldsymbol{a}\times\boldsymbol{b}|^2+(\boldsymbol{a}\cdot\boldsymbol{b})^2=(|\boldsymbol{a}||\boldsymbol{b}|\sin\theta)^2+(|\boldsymbol{a}||\boldsymbol{b}|\cos\theta)^2=|\boldsymbol{a}|^2|\boldsymbol{b}|^2$.

(b) $\boldsymbol{a}\times(\boldsymbol{b}\times(\boldsymbol{a}\times\boldsymbol{b}))=\boldsymbol{a}\times((\boldsymbol{b}\cdot\boldsymbol{b})\boldsymbol{a}-(\boldsymbol{b}\cdot\boldsymbol{a})\boldsymbol{b})=-\boldsymbol{a}\cdot\boldsymbol{b}\,\boldsymbol{a}\times\boldsymbol{b}$.

(c) $(\boldsymbol{a}-\boldsymbol{b})\cdot(\boldsymbol{b}-\boldsymbol{c})\times(\boldsymbol{c}-\boldsymbol{a})=(\boldsymbol{a}-\boldsymbol{b})\cdot(\boldsymbol{b}\times\boldsymbol{c}-\boldsymbol{b}\times\boldsymbol{a}+\boldsymbol{c}\times\boldsymbol{a})=\boldsymbol{a}\cdot\boldsymbol{b}\times\boldsymbol{c}-\boldsymbol{b}\cdot\boldsymbol{c}\times\boldsymbol{a}=0$.

(d) First consider $(\boldsymbol{b}\times\boldsymbol{c})\times(\boldsymbol{c}\times\boldsymbol{a})$. This can be written $\boldsymbol{d}\times(\boldsymbol{c}\times\boldsymbol{a})$, where $\boldsymbol{d}=\boldsymbol{b}\times\boldsymbol{c}$, which using (1.9) is $(\boldsymbol{d}\cdot\boldsymbol{a})\boldsymbol{c}-(\boldsymbol{d}\cdot\boldsymbol{c})\boldsymbol{a}=(\boldsymbol{b}\times\boldsymbol{c}\cdot\boldsymbol{a})\boldsymbol{c}$ since $\boldsymbol{d}\cdot\boldsymbol{c}=0$. Now take the dot product with $(\boldsymbol{a}\times\boldsymbol{b})$, which gives the result $((\boldsymbol{a}\times\boldsymbol{b})\cdot\boldsymbol{c})^2$.

1.12 Take the cross product of the second equation with $\boldsymbol{a}$. Using (1.9) this can be written $|\boldsymbol{a}|^2\boldsymbol{x}-\boldsymbol{a}\cdot\boldsymbol{x}\,\boldsymbol{a}=\boldsymbol{a}\times\boldsymbol{b}$. Now using the first equation this simplifies to $|\boldsymbol{a}|^2\boldsymbol{x}-\boldsymbol{a}=\boldsymbol{a}\times\boldsymbol{b}$, which can be rearranged to solve for $\boldsymbol{x}$: $\boldsymbol{x}=(\boldsymbol{a}\times\boldsymbol{b}+\boldsymbol{a})/|\boldsymbol{a}|^2$. Geometrically, the two equations given in the question represent the equations of a plane and a line. The solution therefore is the point at which the line and plane meet.

1.13 Multiply the first equation by $\boldsymbol{b}$ and the second by $\boldsymbol{a}$, giving $\boldsymbol{r}\cdot\boldsymbol{a}\,\boldsymbol{b}=\boldsymbol{b}$ and $\boldsymbol{r}\cdot\boldsymbol{b}\,\boldsymbol{a}=\boldsymbol{a}$. Subtracting these two equations gives $\boldsymbol{r}\cdot\boldsymbol{a}\,\boldsymbol{b}-\boldsymbol{r}\cdot\boldsymbol{b}\,\boldsymbol{a}=\boldsymbol{b}-\boldsymbol{a}$, which can be written $\boldsymbol{r}\times(\boldsymbol{b}\times\boldsymbol{a})=\boldsymbol{b}-\boldsymbol{a}$. This is now in the form of the equation for a straight line.

1.14 (a) Set $\boldsymbol{a}\times\boldsymbol{b}=\alpha\boldsymbol{a}+\beta\boldsymbol{b}+\gamma\boldsymbol{c}$. The coefficient α is found by taking the dot product with $\boldsymbol{b}\times\boldsymbol{c}$, and similarly for β and γ, giving

$$\begin{aligned}\alpha&=(\boldsymbol{a}\cdot\boldsymbol{b}\boldsymbol{b}\cdot\boldsymbol{c}-\boldsymbol{a}\cdot\boldsymbol{c}|\boldsymbol{b}|^2)/\boldsymbol{a}\cdot\boldsymbol{b}\times\boldsymbol{c},\\ \beta&=(\boldsymbol{a}\cdot\boldsymbol{c}\boldsymbol{b}\cdot\boldsymbol{a}-\boldsymbol{b}\cdot\boldsymbol{c}|\boldsymbol{a}|^2)/\boldsymbol{a}\cdot\boldsymbol{b}\times\boldsymbol{c},\\ \gamma&=(|\boldsymbol{a}|^2|\boldsymbol{b}|^2-(\boldsymbol{a}\cdot\boldsymbol{b})^2)/\boldsymbol{a}\cdot\boldsymbol{b}\times\boldsymbol{c}.\end{aligned}$$

(b) Dot the equation obtained above with $\boldsymbol{c}$ and multiply through by $\boldsymbol{a}\times\boldsymbol{b}\cdot\boldsymbol{c}$:

$$\begin{aligned}(\boldsymbol{a}\times\boldsymbol{b}\cdot\boldsymbol{c})^2&=(\boldsymbol{a}\cdot\boldsymbol{b}\boldsymbol{b}\cdot\boldsymbol{c}-\boldsymbol{a}\cdot\boldsymbol{c}|\boldsymbol{b}|^2)\boldsymbol{a}\cdot\boldsymbol{c}\\ &\quad+(\boldsymbol{a}\cdot\boldsymbol{c}\boldsymbol{b}\cdot\boldsymbol{a}-\boldsymbol{b}\cdot\boldsymbol{c}|\boldsymbol{a}|^2)\boldsymbol{b}\cdot\boldsymbol{c}+(|\boldsymbol{a}|^2|\boldsymbol{b}|^2-(\boldsymbol{a}\cdot\boldsymbol{b})^2)|\boldsymbol{c}|^2\\ &=|\boldsymbol{a}|^2|\boldsymbol{b}|^2|\boldsymbol{c}|^2-(\boldsymbol{a}\cdot\boldsymbol{c})^2|\boldsymbol{b}|^2-(\boldsymbol{a}\cdot\boldsymbol{b})^2|\boldsymbol{c}|^2-(\boldsymbol{b}\cdot\boldsymbol{c})^2|\boldsymbol{a}|^2\\ &\quad+2(\boldsymbol{a}\cdot\boldsymbol{b})(\boldsymbol{b}\cdot\boldsymbol{c})(\boldsymbol{a}\cdot\boldsymbol{c}).\end{aligned}$$

(c) Since the faces of the tetrahedron are equilateral triangles, the dot product of any two vectors forming the edges is 1/2. The volume V is $|(\boldsymbol{a}\times\boldsymbol{b}\cdot\boldsymbol{c})|/6$, so $36V^2=1-1/4-1/4-1/4+2/8=1/2$ and $V=1/6\sqrt{2}=\sqrt{2}/12$.

1.15 The rate of change of $\boldsymbol{h}$ is

$$\frac{\partial\boldsymbol{h}}{\partial t}=m\frac{\partial\boldsymbol{r}}{\partial t}\times\boldsymbol{v}+m\boldsymbol{r}\times\frac{\partial\boldsymbol{v}}{\partial t}.$$

The first term is zero since $\partial \boldsymbol{r}/\partial t = \boldsymbol{v}$ and $\boldsymbol{v}\times\boldsymbol{v} = \boldsymbol{0}$. The second term is also zero since $\partial \boldsymbol{v}/\partial t$ is the acceleration of the particle, which from Newton's second law is proportional to the force $\boldsymbol{F}$ which is proportional to $\boldsymbol{r}$.

1.16 The contour lines are $x^2 - y =$ constant, i.e. $y = x^2 +$ constant. These are parabolas, shifted by different amounts in the y direction, as shown below.

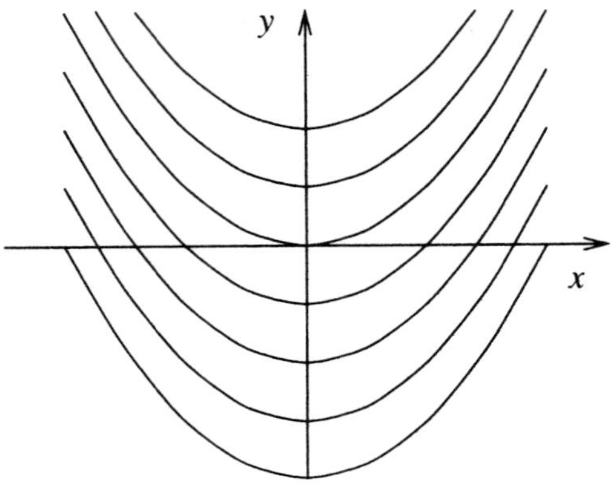

1.17 At the point $(0, 1)$, $\boldsymbol{u}(x, y) = (x + y, -x) = (1, 0)$. At any other point on the y-axis, $x = 0$ so the y component of $\boldsymbol{u}$ is zero and the vector field points horizontally, to the right if $y > 0$ and to the left if $y < 0$. Similarly, on the line $x + y = 0$ the vector field points vertically. By considering a few other points the vector field can be sketched, as shown in the following figure.

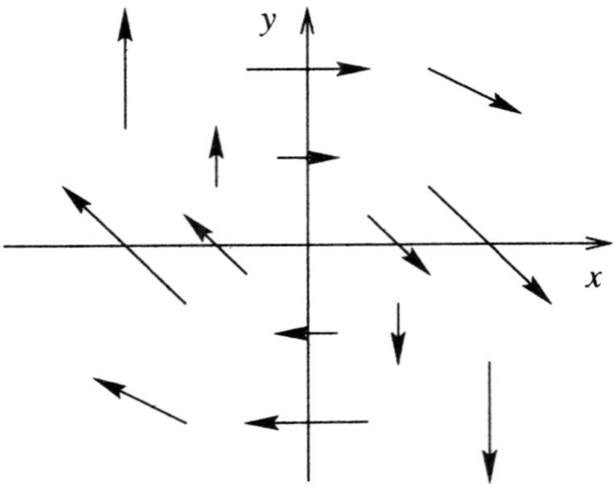

Solutions to Exercises for Chapter 2

2.1 The parametric form of the curve is $x = t$, $y = t^2$, $z = t^2$, so $\boldsymbol{dr} = (dx, dy, dz) = (1, 2t, 2t)\,dt$ and in terms of the parameter t the vector field $\boldsymbol{F}$ is $\boldsymbol{F} = (5t^4, 2t, t + 2t^2)$. So $\boldsymbol{F} \cdot \boldsymbol{dr} = 5t^4 + 6t^2 + 4t^3$ and the value of the integral is

$$\int_C \boldsymbol{F} \cdot \boldsymbol{dr} = \int_0^1 5t^4 + 6t^2 + 4t^3\,dt = 4.$$

2.2 First the parametric form of the line joining the points $(0, 0, 0)$ and $(1, 1, 1)$ must be found. The general form of the equation of a line is $\boldsymbol{r} = \boldsymbol{a} + t\boldsymbol{u}$ (1.6). Choosing $t = 0$ to correspond to $(0, 0, 0)$ gives $\boldsymbol{a} = \boldsymbol{0}$ and choosing $t = 1$ to correspond to $(1, 1, 1)$ gives $\boldsymbol{u} = (1, 1, 1)$, so the equation of the line is $\boldsymbol{r} = t(1, 1, 1)$ and hence $\boldsymbol{dr} = dt(1, 1, 1)$. The vector field $\boldsymbol{F} = (5z^2, 2x, x + 2y)$ can be written in terms of t as $(5t^2, 2t, 3t)$, so $\boldsymbol{F} \cdot \boldsymbol{dr} = 5t^2 + 5t\,dt$ and the integral is

$$\int_C \boldsymbol{F} \cdot \boldsymbol{dr} = \int_0^1 5t^2 + 5t\,dt = 25/6.$$

Notice that the answer is different from the previous answer, even though the same vector field was used and the start and end points of the curve C are the same in each case. In other words, for this vector field the value of the line integral depends on the path chosen between the two points. Hence $\boldsymbol{F}$ is not a conservative vector field.

2.3 Using x as the parameter to evaluate the integral, $\boldsymbol{dr} = (1, \mathrm{e}^x, \mathrm{e}^x)\,dx$ and $\boldsymbol{u} = (y^2, x, z) = (\mathrm{e}^{2x}, x, \mathrm{e}^x)$, so $\boldsymbol{u} \cdot \boldsymbol{dr} = (2\mathrm{e}^{2x} + x\mathrm{e}^x)\,dx$. The integral is

$$\int_C \boldsymbol{u} \cdot \boldsymbol{dr} = \int_0^1 2\mathrm{e}^{2x} + x\mathrm{e}^x\,dx = \left[\mathrm{e}^{2x} + x\mathrm{e}^x - \mathrm{e}^x\right]_0^1 = \mathrm{e}^2$$

using integration by parts.

2.4 The parametric form of the ellipse is $x = a\cos\theta$, $y = b\sin\theta$, $z = 0$, $0 \le \theta \le 2\pi$, so $\boldsymbol{dr} = (-a\sin\theta, b\cos\theta, 0)\,d\theta$ and $\boldsymbol{r} \times \boldsymbol{dr} = (0, 0, ab)\,d\theta$. This means that the integral $\oint_C \boldsymbol{r} \times \boldsymbol{dr}$ only has a component in the z direction and the magnitude of the integral is $2\pi ab$. Note that this is twice the area of the ellipse.

2.5 Since the surface is $z = 0$, the normal is $\boldsymbol{n} = (0, 0, 1)$. Hence $\boldsymbol{u} \cdot \boldsymbol{n} = x + y$ and the integral is

$$\begin{aligned}\iint_S \boldsymbol{u} \cdot \boldsymbol{n}\,dS &= \int_0^1 \int_0^2 x + y\,dy\,dx = \int_0^1 \left[xy + y^2/2\right]_0^2 dx \\ &= \int_0^1 2x + 2\,dx = \left[x^2 + 2x\right]_0^1 = 3.\end{aligned}$$

Note that the y integral was carried out first. However, it is equally possible to carry out the x integral first, and this gives the same result.

2.6 This surface integral has six parts, corresponding to the six faces of the cube shown below, which must be considered separately. On the surface

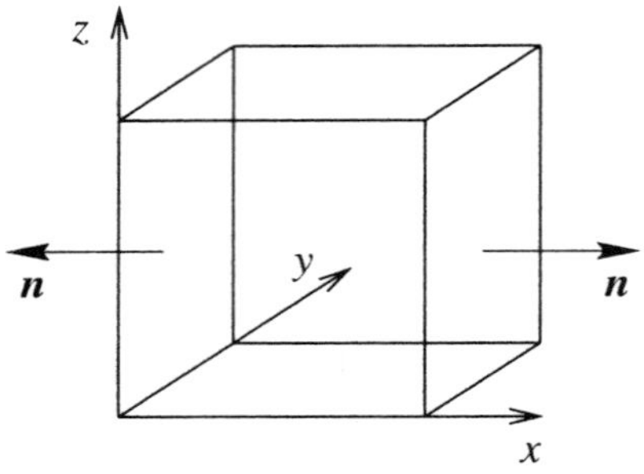

$x = 0$, $\boldsymbol{n} = (-1, 0, 0)$ and $\boldsymbol{u} = \boldsymbol{r} = (x, y, z) = (0, y, z)$ so $\boldsymbol{u} \cdot \boldsymbol{n} = 0$ and there is no contribution to the integral from this surface. By symmetry the same result holds for the surfaces $y = 0$ and $z = 0$. On the surface $x = 1$, $\boldsymbol{n} = (1, 0, 0)$ and $\boldsymbol{u} \cdot \boldsymbol{n} = x = 1$. Since this is a constant, the integral over this surface is just the value of the constant multiplied by the area of the surface, which gives 1. The same result holds on the surfaces $y = 1$ and $z = 1$, so the total value of the surface integral, adding the six contributions, is $0 + 0 + 0 + 1 + 1 + 1 = 3$.

2.7 The two curves $y = x^2$ and $x = y^2$ meet at $(0, 0)$ and $(1, 1)$. Doing the x integral first, the limits are $y^2 \le x \le \sqrt{y}$ and $0 \le y \le 1$, so

$$\iint_S \boldsymbol{u} \cdot \boldsymbol{n}\, dS = \int_0^1 \int_{y^2}^{\sqrt{y}} x^2\, dx\, dy = \int_0^1 (y^{3/2} - y^6)/3\, dy = 3/35.$$

2.8 The surface is written parametrically as $(x, y, x+y^2)$, so two vectors parallel to the surface are $(1, 0, 1)$ and $(0, 1, 2y)$. By taking the cross product of these two vectors, $\boldsymbol{n}\, dS = (-1, -2y, 1)\, dx\, dy$. Since this has positive z component the direction of $\boldsymbol{n}$ must be changed, so $\boldsymbol{n}\, dS = (1, 2y, -1)\, dx\, dy$ and $\boldsymbol{u} \cdot \boldsymbol{n}\, dS = x\, dx\, dy$. In terms of x and y the region of integration is $x + y^2 < 0$, $x > -1$, so doing the x integral first, $-1 < x < -y^2$ and $-1 < y < 1$, and the integral is

$$\iint_S \boldsymbol{u} \cdot \boldsymbol{n}\, dS = \int_{-1}^1 \int_{-1}^{-y^2} x\, dx\, dy = \int_{-1}^1 y^4/2 - 1/2\, dy = 1/5 - 1 = -4/5.$$

2.9 The required volume integral is

$$\begin{aligned}\iiint_V \phi\, dV &= \int_0^1 \int_1^2 \int_0^3 x^2 + y^2 + z^2\, dz\, dy\, dx \\ &= \int_0^1 \int_1^2 3x^2 + 3y^2 + 9\, dy\, dx \\ &= \int_0^1 \left[3x^2 y + y^3 + 9y\right]_1^2 dx \\ &= \int_0^1 3x^2 + 16\, dx = 17.\end{aligned}$$

2.10 Doing the integrals in the order z, y, x, the range of z between the two planes is $-x-1 < z < x+1$ and y ranges from $-\sqrt{1-x^2}$ to $\sqrt{1-x^2}$ at a fixed value of x. Finally, the limits on the outer x integral are -1 and 1. The volume is

$$\begin{aligned}\int_{-1}^1 \int_{-\sqrt{1-x^2}}^{\sqrt{1-x^2}} \int_{-x-1}^{x+1} dz\, dy\, dx &= \int_{-1}^1 \int_{-\sqrt{1-x^2}}^{\sqrt{1-x^2}} 2(x+1)\, dy\, dx \\ &= \int_{-1}^1 4\sqrt{1-x^2}\,(x+1)\, dx = 2\pi.\end{aligned}$$

2.11 The edge of the pond is where $z = 0$, so $x^2 + y^2 = 1$. The limits of the volume integral are obtained as follows. Doing the z integral first, at a fixed value of x and y, z ranges from 0 to $1 - x^2 - y^2$. The limits on x and y are the same as in the previous exercise, so the volume V is

$$\begin{aligned}V &= \iiint_V dV = \int_{-1}^1 \int_{-\sqrt{1-x^2}}^{\sqrt{1-x^2}} \int_0^{1-x^2-y^2} dz\, dy\, dx \\ &= \int_{-1}^1 \int_{-\sqrt{1-x^2}}^{\sqrt{1-x^2}} 1 - x^2 - y^2\, dy\, dx \\ &= \int_{-1}^1 \left[y - x^2 y - y^3/3\right]_{-\sqrt{1-x^2}}^{\sqrt{1-x^2}} dx \\ &= \frac{4}{3}\int_{-1}^1 (1-x^2)\sqrt{1-x^2}\, dx = \pi/2,\end{aligned}$$

where the results of Section 2.1.2 have been used to find the last integral. The volume of the pond is therefore approximately 1.57 m^3.
In the case of a hemisphere, the volume is $2\pi/3$, so the volume of the hemisphere is greater by a factor 4/3.

Solutions to Exercises for Chapter 3

3.1 $f = xyz$, so $\nabla f = (yz, xz, xy)$. At the point $(1,2,3)$ this has the value $(6,3,2)$. The directional derivative in the direction of the vector $(1,1,0)$ is found by taking the dot product of ∇f and the unit vector in this direction. This is $(6,3,2) \cdot (1,1,0)/\sqrt{2} = 9/\sqrt{2}$.

3.2 First write the equation of the surface in the form $f =$ constant, so $f = y - x - z^3 = 0$. A vector normal to the surface is $\nabla f = (-1, 1, -3z^2)$. At the point $(1,2,1)$ this is $\nabla f = (-1,1,-3)$. To find the unit normal, divide by the magnitude, so $\boldsymbol{n} = (-1,1,-3)/\sqrt{11}$. The normal pointing in the opposite direction, $(1,-1,3)/\sqrt{11}$, is an equally valid answer.

3.3 $\phi = r = |\boldsymbol{r}| = (x^2+y^2+z^2)^{1/2}$, so

$$\frac{\partial \phi}{\partial x} = \frac{1}{2}(x^2+y^2+z^2)^{-1/2}\,2x = \frac{x}{(x^2+y^2+z^2)^{1/2}}.$$

Similarly,

$$\frac{\partial \phi}{\partial y} = \frac{y}{(x^2+y^2+z^2)^{1/2}}, \quad \frac{\partial \phi}{\partial z} = \frac{z}{(x^2+y^2+z^2)^{1/2}}.$$

Hence $\nabla \phi = (x,y,z)(x^2+y^2+z^2)^{-1/2}$ which can also be written as $\boldsymbol{r}/r$ or as $\hat{r}$, the unit vector in the direction of $\boldsymbol{r}$.

Geometrically, the level surfaces $\phi =$ constant are concentric spheres centred at the origin. The vector $\nabla \phi$ points in a direction perpendicular to these surfaces, i.e. radially away from the origin.

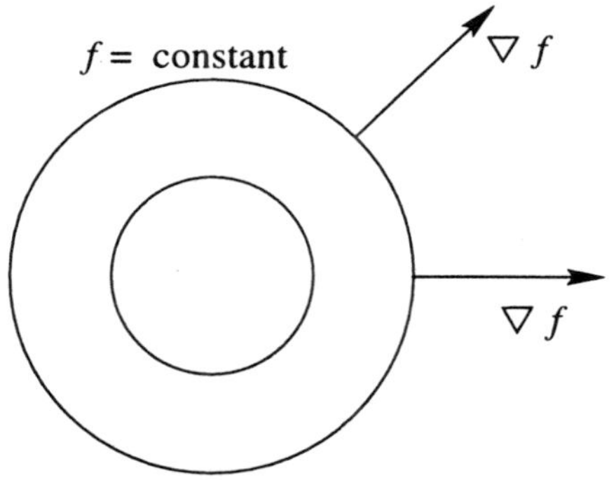

3.4 The sphere and the cylinder intersect when $x^2+y^2+z^2=2$ and $x^2+y^2=1$. Subtracting these two equations gives $z^2=1$, so the points of intersection are the two circles $x^2+y^2=1$, $z=\pm 1$.
The angle between the two surfaces is the angle between the normals to the surfaces. By taking the gradients of the two surfaces, these normal vectors are $\boldsymbol{n}_1=(2x,2y,2z)$ and $\boldsymbol{n}_2=(2x,2y,0)$. The angle θ between the normals is found using the dot product, $\boldsymbol{n}_1\cdot\boldsymbol{n}_2=|\boldsymbol{n}_1||\boldsymbol{n}_2|\cos\theta$. This gives

$$4x^2+4y^2=2\sqrt{x^2+y^2+z^2}\,2\sqrt{x^2+y^2}\cos\theta.$$

At a point of intersection this simplifies to $4=2\sqrt{2}\,2\cos\theta$, so $\cos\theta=1/\sqrt{2}$ and hence $\theta=45°$. Note that this is the same for all the points of intersection.

3.5 $\boldsymbol{\nabla} f=(2xy, x^2+3y^2-1)$. This is zero when $xy=0$, so either x or y must be zero. Thus f has maxima, minima or saddle points at the points $(\pm 1,0)$, where $f=0$; $(0,1/\sqrt{3})$, where $f=-2/3\sqrt{3}$ and $(0,-1/\sqrt{3})$, where $f=2/3\sqrt{3}$. Since $f=y(x^2+y^2-1)$, the contour $f=0$ includes the line $y=0$ and the circle $x^2+y^2=1$. This means that the points at $(\pm 1,0)$ must be saddle points. Putting together all this information, the sketch of f and its gradient is as shown below. Lines are contours $f=$ constant and arrows are $\boldsymbol{\nabla} f$.

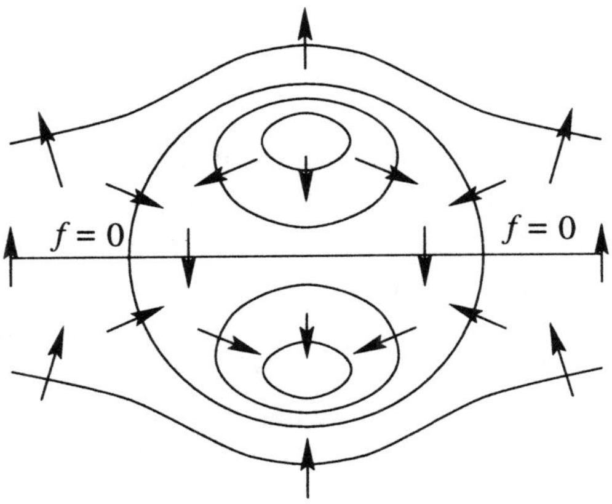

3.6 $f=\boldsymbol{a}\cdot\boldsymbol{r}=a_1x+a_2y+a_3z$, so $\boldsymbol{\nabla} f=(a_1,a_2,a_3)=\boldsymbol{a}$. Geometrically, $f=$ constant is the equation of a plane (1.3) which is perpendicular to the vector $\boldsymbol{a}$.

3.7 Proceeding as in Example 3.6, the vector field $\boldsymbol{F}=(\sin y, x, 0)$ is conservative if and only if there is a function ϕ satisfying $\boldsymbol{F}=\boldsymbol{\nabla}\phi$, so

$$\frac{\partial\phi}{\partial x} = \sin y, \qquad \frac{\partial\phi}{\partial y} = x, \qquad \frac{\partial\phi}{\partial z} = 0.$$

The first equation gives $\phi = x\sin y + h(y,z)$, where h is an arbitrary function of y and z. The second equation then becomes

$$x\cos y + \frac{\partial h}{\partial y} = x.$$

This equation cannot be satisfied for all values of x, since the terms involving x do not match and h does not depend on x. Hence $\boldsymbol{F}$ is not a conservative vector field.

3.8 If $\boldsymbol{F}$ can be written as $\boldsymbol{\nabla}\phi$ then

$$\frac{\partial\phi}{\partial x} = y/(x^2+y^2), \qquad \frac{\partial\phi}{\partial y} = -x/(x^2+y^2), \qquad \frac{\partial\phi}{\partial z} = 0.$$

Integrating the first of these equations (using the substitution $x = y\tan\theta$) gives $\phi = \tan^{-1}(x/y) + h(y,z)$ for any function h. From the second and third equations it follows that h may be taken to be zero, so $\boldsymbol{F} = \boldsymbol{\nabla}\phi$ where $\phi = \tan^{-1}(x/y)$.

Now consider the line integral of $\boldsymbol{F}$ around the unit circle $x^2 + y^2 = 1$, $z = 0$, given parametrically by $x = \cos\theta$, $y = \sin\theta$, $z = 0$, $0 \le \theta \le 2\pi$.

$$\begin{aligned}\oint_C \boldsymbol{F}\cdot d\boldsymbol{r} &= \int_0^{2\pi} (\sin\theta, -\cos\theta, 0)\cdot(-\sin\theta, \cos\theta, 0)\, d\theta \\ &= \int_0^{2\pi} -1\, d\theta = -2\pi.\end{aligned}$$

At first sight this appears to contradict Theorem 3.1, since we have a non-conservative vector field which can be written as the gradient of a potential. The resolution of the conflict is that both $\boldsymbol{F}$ and ϕ are undefined at the origin, so Theorem 3.1 does not apply.

3.9 The components of $\boldsymbol{\nabla}\phi$ are just the partial derivatives of ϕ with respect to x, y and z, so for $\phi = x^2 + xy + yz^2$, $\boldsymbol{\nabla}\phi = (2x+y, x+z^2, 2yz)$. The Laplacian can be found either by taking the divergence of this vector, or the sum of the second partial derivatives of ϕ, giving the result $\nabla^2\phi = 2 + 2y$.

3.10 The gradient $\boldsymbol{\nabla}\phi$ is

$$\begin{aligned}\boldsymbol{\nabla}\phi = \Big(& k\cos(kx)\sin(ly)\exp(\sqrt{k^2+l^2}z), \\ & l\sin(kx)\cos(ly)\exp(\sqrt{k^2+l^2}z), \\ & \sqrt{k^2+l^2}\sin(kx)\sin(ly)\exp(\sqrt{k^2+l^2}z)\Big).\end{aligned}$$

Now take the divergence of this vector:

$$\begin{aligned}\nabla^2\phi &= -k^2\sin(kx)\sin(ly)\exp(\sqrt{k^2+l^2}z)\\ &\quad -l^2\sin(kx)\sin(ly)\exp(\sqrt{k^2+l^2}z)\\ &\quad +(k^2+l^2)\sin(kx)\sin(ly)\exp(\sqrt{k^2+l^2}z)\\ &= 0.\end{aligned}$$

This means that the function ϕ is a solution to the equation $\nabla^2\phi = 0$, which is Laplace's equation.

3.11 The unit normal to the surface ϕ = constant is $\boldsymbol{\nabla}\phi/|\boldsymbol{\nabla}\phi|$. Here, $\phi = xy^2 + 2yz$ so $\boldsymbol{\nabla}\phi = (y^2, 2xy+2z, 2y)$. At the point $(-2,2,3)$, $\boldsymbol{\nabla}\phi = (4,-2,4)$ which has magnitude $\sqrt{16+4+16} = 6$, so the unit normal is $\boldsymbol{n} = (2/3, -1/3, 2/3)$.

3.12 For $\phi(x,y,z) = x^2+y^2+z^2+xy-3x$, $\boldsymbol{\nabla}\phi = (2x+y-3, 2y+x, 2z)$. ϕ has a minimum or maximum where $\boldsymbol{\nabla}\phi = \boldsymbol{0}$, which gives $2x+y=3$, $2y=-x$, $z=0$, so $x=2$, $y=-1$, $z=0$. At this point the value of ϕ is -3. Since ϕ becomes large and positive when x, y or z become large, this must be a minimum value.

3.13 First find a normal to the surface. If $f = x^2+y^2-2z^3$ then $\boldsymbol{\nabla}f = (2x, 2y, -6z^2)$ so a normal to the surface at the point $(1,1,1)$ is $(2,2,-6)$. The equation of the plane is therefore $2x+2y-6z$ = constant, using (1.3). Imposing that the plane must pass through $(1,1,1)$ gives the value of the constant to be -2.

3.14 (a) For $\boldsymbol{u} = (y,z,x)$,

$$\boldsymbol{\nabla}\cdot\boldsymbol{u} = \frac{\partial y}{\partial x} + \frac{\partial z}{\partial y} + \frac{\partial x}{\partial z} = 0.$$

$$\boldsymbol{\nabla}\times\boldsymbol{u} = \left(\frac{\partial x}{\partial y} - \frac{\partial z}{\partial z}, \frac{\partial y}{\partial z} - \frac{\partial x}{\partial x}, \frac{\partial z}{\partial x} - \frac{\partial y}{\partial y}\right) = (-1,-1,-1).$$

(b) For $\boldsymbol{v} = (xyz, z^2, x-y)$,

$$\boldsymbol{\nabla}\cdot\boldsymbol{v} = \frac{\partial(xyz)}{\partial x} + \frac{\partial(z^2)}{\partial y} + \frac{\partial(x-y)}{\partial z} = yz + 0 + 0 = yz.$$

$$\begin{aligned}\boldsymbol{\nabla}\times\boldsymbol{v} &= \left[\frac{\partial(x-y)}{\partial y} - \frac{\partial(z^2)}{\partial z}, \frac{\partial(xyz)}{\partial z} - \frac{\partial(x-y)}{\partial x}, \frac{\partial(z^2)}{\partial x} - \frac{\partial(xyz)}{\partial y}\right]\\ &= (-1-2z, xy-1, -xz).\end{aligned}$$

3.15 These results follow directly from the linearity of the differential operator:

$$\begin{aligned}\boldsymbol{\nabla}\cdot(c\boldsymbol{u}+d\boldsymbol{v}) &= \frac{\partial}{\partial x}(cu_1+dv_1) + \frac{\partial}{\partial y}(cu_2+dv_2) + \frac{\partial}{\partial z}(cu_3+dv_3)\\ &= c\frac{\partial u_1}{\partial x} + d\frac{\partial v_1}{\partial x} + c\frac{\partial u_2}{\partial y} + d\frac{\partial v_2}{\partial y} + c\frac{\partial u_3}{\partial z} + d\frac{\partial v_3}{\partial z}\\ &= c\boldsymbol{\nabla}\cdot\boldsymbol{u} + d\boldsymbol{\nabla}\cdot\boldsymbol{v},\end{aligned}$$

and similarly for curl.

3.16 $\boldsymbol{u}$ is irrotational if $\nabla \times \boldsymbol{u} = \boldsymbol{0}$. $\nabla \times \boldsymbol{u} = (1-1, ax-2x, b\cos x - \cos x)$, so the solution is $a = 2$, $b = 1$.

3.17 (a) $\nabla \times \boldsymbol{u} = (-2z\sin y + 2yx + 2z\sin y - 2xy, y^2 - y^2, 2yz - 2yz) = \boldsymbol{0}$ so $\boldsymbol{u}$ is irrotational.

(b) The potential function ϕ for which $\boldsymbol{u} = \nabla\phi$ can be found using the step-by-step method of Example 3.6 or as follows: $\partial\phi/\partial x = y^2 z$ suggests $\phi = xy^2z$. Taking $\nabla\phi$ gives all the terms in $\boldsymbol{u}$ except the trigonometric terms. The z component suggests $\phi = z^2\cos y + xy^2z$. Taking the gradient of ϕ gives all the terms in $\boldsymbol{u}$, so this (plus an arbitrary constant) is the potential function.

(c) The line integral of $\boldsymbol{u}$ along the curve is just the difference between the values of ϕ at the endpoints. The endpoints are $(0,0,0)$ and $(1,0,1)$, so the line integral is $\phi(1,0,1) - \phi(0,0,0) = 1$.

Solutions to Exercises for Chapter 4

4.1 $\epsilon_{ijk}a_jb_k + a_jd_jc_i = e_i$.

4.2 First, tidy up and rearrange the equation using (4.3) and the symmetry property $\epsilon_{kji} = \epsilon_{ikj}$: $c_i + \epsilon_{ikj}a_kb_j = d_lb_le_mc_mc_i$. Now this can be written as the vector equation $\boldsymbol{c} + \boldsymbol{a}\times\boldsymbol{b} = (\boldsymbol{d}\cdot\boldsymbol{b})(\boldsymbol{e}\cdot\boldsymbol{c})\boldsymbol{c}$.

4.3 $[\boldsymbol{a}\times\boldsymbol{b}]_i = \epsilon_{ijk}a_jb_k = -\epsilon_{ikj}a_jb_k = -\epsilon_{ikj}b_ka_j = -[\boldsymbol{b}\times\boldsymbol{a}]_i$.

4.4 (a) $\delta_{ij}\epsilon_{ijk}$: this expression is always zero, since if $i = j$ then $\epsilon_{ijk} = 0$, while if $i \neq j$ then $\delta_{ij} = 0$.

(b) $\epsilon_{ijk}\epsilon_{ilm} = \epsilon_{jki}\epsilon_{ilm}$ using (4.6). Using (4.12) this is $\delta_{jl}\delta_{km} - \delta_{jm}\delta_{kl}$.

(c) $\epsilon_{ijk}\epsilon_{ijm} = \delta_{jj}\delta_{km} - \delta_{jm}\delta_{kj}$, using (b) with $l = j$. Using $\delta_{jj} = 3$ and the substitution property of δ_{ij} this can be simplified to $3\delta_{km} - \delta_{km} = 2\delta_{km}$.

(d) Using (c) with $m = k$, $\epsilon_{ijk}\epsilon_{ijk} = 2\delta_{kk} = 6$, as obtained in Example 4.9.

4.5 Using suffix notation,

$$\begin{aligned}
\boldsymbol{a}\times\boldsymbol{b}\cdot\boldsymbol{c}\times\boldsymbol{d} &= \epsilon_{ijk}a_jb_k\epsilon_{ilm}c_ld_m \\
&= (\delta_{jl}\delta_{km} - \delta_{jm}\delta_{kl})a_jb_kc_ld_m \\
&= a_lb_mc_ld_m - a_mb_lc_ld_m \\
&= (\boldsymbol{a}\cdot\boldsymbol{c})(\boldsymbol{b}\cdot\boldsymbol{d}) - (\boldsymbol{a}\cdot\boldsymbol{d})(\boldsymbol{b}\cdot\boldsymbol{c}).
\end{aligned}$$

4.6 Let $C = AB$, so in suffix notation $C_{ij} = A_{ik}B_{kj}$. Then

$$(AB)^T_{ij} = C^T_{ij} = C_{ji} = A_{jk}B_{ki}.$$

Now consider $B^T A^T$:

$$(B^T A^T)_{ij} = B^T_{ik} A^T_{kj} = B_{ki} A_{jk}.$$

These two expressions are the same (recall that ordering of terms does not matter in suffix notation) so $(AB)^T = B^T A^T$.

4.7 The determinant of a 3×3 matrix M is

$$|M| = \begin{vmatrix} M_{11} & M_{12} & M_{13} \\ M_{21} & M_{22} & M_{23} \\ M_{31} & M_{32} & M_{33} \end{vmatrix}.$$

Expanding this determinant,

$$\begin{aligned} |M| &= M_{11}(M_{22}M_{33} - M_{23}M_{32}) + M_{12}(M_{23}M_{31} - M_{21}M_{33}) \\ &\quad + M_{13}(M_{21}M_{32} - M_{22}M_{31}). \end{aligned} \qquad (8.37)$$

Now expand the suffix notation expression $\epsilon_{ijk} M_{1i} M_{2j} M_{3k}$. Since i, j and k are repeated, there is a sum over all three indices, so

$$\epsilon_{ijk} M_{1i} M_{2j} M_{3k} = \sum_{i=1}^{3} \sum_{j=1}^{3} \sum_{k=1}^{3} \epsilon_{ijk} M_{1i} M_{2j} M_{3k}.$$

Since only six of the 27 elements of ϵ_{ijk} are non-zero, there are six terms in this sum, and writing them out gives (8.37), so we have shown that $|M| = \epsilon_{ijk} M_{1i} M_{2j} M_{3k}$.

Now turn to the formula (4.10), $\epsilon_{pqr}|M| = \epsilon_{ijk} M_{pi} M_{qj} M_{rk}$. First note that the formula is true for $p = 1$, $q = 2$ and $r = 3$, since in this case it reduces to the result shown above. Now consider the effect of interchanging p and q. The l.h.s. changes sign, since $\epsilon_{pqr} = -\epsilon_{qpr}$. The r.h.s. becomes

$$\begin{aligned} \epsilon_{ijk} M_{qi} M_{pj} M_{rk} &= \epsilon_{jik} M_{qj} M_{pi} M_{rk} \quad \text{(relabelling } i \leftrightarrow j) \\ &= -\epsilon_{ijk} M_{pi} M_{qj} M_{rk}, \end{aligned}$$

so the r.h.s. also changes sign when p and q are interchanged. Similarly, both sides change sign when any two of p, q and r are interchanged. This suffices to prove the result, since both sides are zero when any two of p, q and r are equal and all permutations of $1, 2, 3$ can be achieved by a suitable sequence of interchanges.

4.8 Make use of $\epsilon_{pqr}|M| = \epsilon_{ijk} M_{pi} M_{qj} M_{rk}$.

(a) Multiplying both sides by ϵ_{pqr} and using the result of Example 4.9, $6|M| = \epsilon_{pqr} \epsilon_{ijk} M_{pi} M_{qj} M_{rk}$. Note that because of the six repeated suffices there are $3^6 = 729$ terms in this sum!

(b) Using the above result,

$$6|M^T| = \epsilon_{pqr}\epsilon_{ijk}M^T_{pi}M^T_{qj}M^T_{rk} = \epsilon_{pqr}\epsilon_{ijk}M_{ip}M_{jq}M_{kr}.$$

Now since i, j, k, p, q and r are all dummy suffices, we can relabel $i \leftrightarrow p$, $j \leftrightarrow q$, $k \leftrightarrow r$, so that the formula for $|M^T|$ is identical to that for $|M|$.
(c) The formulae for $|M|$ and $|N|$ are $\epsilon_{pqr}|M| = \epsilon_{ijk}M_{pi}M_{qj}M_{rk}$, $\epsilon_{pqr}|N| = \epsilon_{lmn}N_{pl}N_{qm}N_{rn}$. By multiplying these together, we obtain

$$6|M||N| = \epsilon_{ijk}\epsilon_{lmn}M_{pi}N_{pl}M_{qj}N_{qm}M_{rk}N_{rn}.$$

Now $M_{pi}N_{pl} = M^T_{ip}N_{pl} = (M^TN)_{il}$, so

$$6|M||N| = \epsilon_{ijk}\epsilon_{lmn}(M^TN)_{il}(M^TN)_{jm}(M^TN)_{kn} = 6|M^TN|,$$

using the result of part (a). Thus we have shown that $|M||N| = |M^TN|$, or equivalently $|M^T||N| = |MN|$. Applying the result of part (b), it follows that $|M||N| = |MN|$.

4.9 Since each term in the equation is a vector, we first introduce a free suffix i for each term: $(\boldsymbol{a} \times \boldsymbol{b})_i + c_i = (\boldsymbol{a} \cdot \boldsymbol{b})b_i - d_i$. Now we introduce dummy suffices for the dot and cross product, making sure that i is not reused: $\epsilon_{ijk}a_jb_k + c_i = a_jb_jb_i - d_i$.

4.10 (a) Using (4.3) twice and (4.4),

$$\delta_{ij}\delta_{jk}\delta_{ki} = \delta_{ik}\delta_{ki} = \delta_{ii} = 3.$$

(b) Using (4.12),

$$\epsilon_{ijk}\epsilon_{klm}\epsilon_{mni} = (\delta_{il}\delta_{jm} - \delta_{im}\delta_{jl})\epsilon_{mni} = \epsilon_{jnl} - \epsilon_{ini}\delta_{jl} = \epsilon_{jnl}.$$

4.11 Using (4.3), $\delta_{ij}a_jb_lc_k\delta_{li} = a_ib_ic_k$. Here the i is a dummy suffix and the k is a free suffix, so the result is the k component of the vector $(\boldsymbol{a} \cdot \boldsymbol{b})\boldsymbol{c}$.

4.12 (a) $\boldsymbol{\nabla} \times (f\boldsymbol{\nabla} f) = \boldsymbol{\nabla} f \times \boldsymbol{\nabla} f + f\boldsymbol{\nabla} \times (\boldsymbol{\nabla} f)$, using (4.28). Each of these terms is zero since any vector crossed with itself gives zero and the combination curl grad is always zero.
(b) $\boldsymbol{\nabla} \cdot (f\boldsymbol{\nabla} f) = \boldsymbol{\nabla} f \cdot \boldsymbol{\nabla} f + f\boldsymbol{\nabla} \cdot (\boldsymbol{\nabla} f)$, from (4.27). This can be simplified to $\boldsymbol{\nabla} \cdot (f\boldsymbol{\nabla} f) = |\boldsymbol{\nabla} f|^2 + f\nabla^2 f$.

4.13 $\boldsymbol{u}$ is solenoidal if its divergence is zero.

$$\begin{aligned} \boldsymbol{\nabla} \cdot \boldsymbol{u} &= \boldsymbol{\nabla} \cdot (\boldsymbol{\nabla} f \times \boldsymbol{\nabla} g) \\ &= (\boldsymbol{\nabla} \times \boldsymbol{\nabla} f) \cdot \boldsymbol{\nabla} g - (\boldsymbol{\nabla} \times \boldsymbol{\nabla} g) \cdot \boldsymbol{\nabla} f \quad \text{using (4.29)} \\ &= 0 \quad \text{using (3.23).} \end{aligned}$$

4.14 Applying (4.35) with $\boldsymbol{u} = \boldsymbol{v}$, the second term is zero and the fifth and sixth terms cancel, leaving

$$\boldsymbol{u} \cdot \boldsymbol{\nabla} \boldsymbol{u} = (\boldsymbol{\nabla}(\boldsymbol{u} \cdot \boldsymbol{u}) - 2\boldsymbol{u} \times (\boldsymbol{\nabla} \times \boldsymbol{u}))/2$$

which is (4.34).

4.15 (a) In suffix notation,

$$\boldsymbol{\nabla}\cdot\nabla^2\boldsymbol{u} = \frac{\partial}{\partial x_i}\nabla^2 u_i = \frac{\partial}{\partial x_i}\frac{\partial^2 u_i}{\partial x_j \partial x_j}.$$

Similarly,

$$\nabla^2\boldsymbol{\nabla}\cdot\boldsymbol{u} = \frac{\partial^2 \boldsymbol{\nabla}\cdot\boldsymbol{u}}{\partial x_j \partial x_j} = \frac{\partial^2}{\partial x_j \partial x_j}\frac{\partial u_i}{\partial x_i},$$

so the two expressions are equal since the order of partial derivatives can be interchanged.

(b) Using (4.24),

$$\begin{aligned}\boldsymbol{\nabla}\cdot\nabla^2\boldsymbol{u} &= \boldsymbol{\nabla}\cdot(\boldsymbol{\nabla}(\boldsymbol{\nabla}\cdot\boldsymbol{u}) - \boldsymbol{\nabla}\times(\boldsymbol{\nabla}\times\boldsymbol{u}))\\ &= \boldsymbol{\nabla}\cdot(\boldsymbol{\nabla}(\boldsymbol{\nabla}\cdot\boldsymbol{u})) \quad \text{(since div curl is zero)}\\ &= \nabla^2(\boldsymbol{\nabla}\cdot\boldsymbol{u}) \quad (\text{since } \nabla^2 = \boldsymbol{\nabla}\cdot\boldsymbol{\nabla}).\end{aligned}$$

Note that the first ∇^2 acts on a vector but the second acts on a scalar, so they must be interpreted differently.

4.16 Take the divergence of the equation:

$$\boldsymbol{\nabla}\cdot\boldsymbol{u} + \boldsymbol{\nabla}\cdot\boldsymbol{\nabla}\times\boldsymbol{w} = \boldsymbol{\nabla}\cdot\boldsymbol{\nabla}\phi + \boldsymbol{\nabla}\cdot\nabla^2\boldsymbol{u}.$$

The first term is zero as $\boldsymbol{u}$ is solenoidal. The second term is zero because the combination div curl is always zero. The last term is also zero since from the previous exercise, $\boldsymbol{\nabla}\cdot\nabla^2\boldsymbol{u} = \nabla^2\boldsymbol{\nabla}\cdot\boldsymbol{u} = 0$. So the equation reduces to $0 = \boldsymbol{\nabla}\cdot\boldsymbol{\nabla}\phi = \nabla^2\phi$ which is Laplace's equation.

4.17

$$[\boldsymbol{\nabla} f(r)]_i = \frac{\partial f(r)}{\partial x_i} = \frac{df(r)}{dr}\frac{\partial r}{\partial x_i} = f'(r)\frac{x_i}{r},$$

using the usual rule for differentiating a function of a function together with the result $\frac{\partial r}{\partial x_i} = x_i/r$ from (4.19). Thus $\boldsymbol{\nabla} f(r) = f'(r)\boldsymbol{r}/r$.

4.18 $\boldsymbol{u} = h(r)\boldsymbol{r}$.

(a) $\boldsymbol{\nabla}\times\boldsymbol{u} = \boldsymbol{\nabla}\times(h(r)\boldsymbol{r}) = \boldsymbol{\nabla}h\times\boldsymbol{r} + h\boldsymbol{\nabla}\times\boldsymbol{r} = h'(r)\boldsymbol{r}\times\boldsymbol{r}/r = \boldsymbol{0}$, using the results of the previous exercise for $\boldsymbol{\nabla}h$.

(b) $\boldsymbol{\nabla}\cdot\boldsymbol{u} = \boldsymbol{\nabla}\cdot(h(r)\boldsymbol{r}) = \boldsymbol{\nabla}h\cdot\boldsymbol{r} + h\boldsymbol{\nabla}\cdot\boldsymbol{r} = h'(r)\boldsymbol{r}\cdot\boldsymbol{r}/r + 3h = rh'(r) + 3h$. So if $\boldsymbol{\nabla}\cdot\boldsymbol{u} = 0$, $h(r)$ obeys the differential equation

$$r\frac{dh}{dr} + 3h = 0.$$

(c) Using the method of separation of variables,

$$\int\frac{dh}{h} = -3\int\frac{dr}{r}$$

which gives $\log h = -3\log r + c = \log(r^{-3}) + c$ for some constant c. Taking the exponential of both sides, $h = A/r^3$ where the constant $A = \exp c$.

4.19 (a) For a Beltrami field, $\nabla \cdot \boldsymbol{u} = \nabla \cdot (c\nabla \times \boldsymbol{u}) = c\nabla \cdot (\nabla \times \boldsymbol{u}) = 0$.
(b) Let $\boldsymbol{v} = \nabla \times \boldsymbol{u}$, so $\boldsymbol{u} = c\boldsymbol{v}$. Taking the curl of this equation, $\nabla \times \boldsymbol{u} = \nabla \times (c\boldsymbol{v})$ so $\boldsymbol{v} = c\nabla \times \boldsymbol{v}$.
(c) If $\boldsymbol{u} = (\sin y, f, g)$, then $\boldsymbol{u} = c\nabla \times \boldsymbol{u}$ gives the three equations

$$\sin y = c\left(\frac{\partial g}{\partial y} - \frac{\partial f}{\partial z}\right), \qquad f = c\left(-\frac{\partial g}{\partial x}\right), \qquad g = c\left(\frac{\partial f}{\partial x} - \cos y\right).$$

Given that g is independent of x, it follows that $f = 0$, $g = -c\cos y$ and $\sin y = c(c\sin y)$. Hence either $c = 1$, $g = -\cos y$ or $c = -1$, $g = \cos y$.

Solutions to Exercises for Chapter 5

5.1 The surface integral is equal to the volume integral of $\nabla \cdot \boldsymbol{u}$, but $\nabla \cdot \boldsymbol{u} = \sin y + 0 - \sin y = 0$, so the value of the integral is zero.

5.2 $\boldsymbol{u} = (y, x, z - x)$, so $\nabla \cdot \boldsymbol{u} = 1$. The volume integral is therefore

$$\iiint_V \nabla \cdot \boldsymbol{u}\, dV = \int_0^1 \int_0^1 \int_0^1 1\, dx\, dy\, dz = 1.$$

The surface integral has six parts from the six faces of the cube. On the face where $x = 0$, $\boldsymbol{n} = (-1, 0, 0)$ and so $\boldsymbol{u} \cdot \boldsymbol{n} = -y$. Similarly, on the face where $x = 1$, $\boldsymbol{n} = (1, 0, 0)$ and $\boldsymbol{u} \cdot \boldsymbol{n} = y$, so the surface integrals from these two faces cancel. The same argument holds for the faces $y = 0$ and $y = 1$. On $z = 0$, $\boldsymbol{n} = (0, 0, -1)$ and $\boldsymbol{u} \cdot \boldsymbol{n} = -z + x = x$, while on $z = 1$, $\boldsymbol{n} = (0, 0, 1)$ and $\boldsymbol{u} \cdot \boldsymbol{n} = z - x = 1 - x$. The integrals over these two surfaces then give

$$\int_0^1 \int_0^1 x + 1 - x\, dx\, dy = 1.$$

Therefore, both the surface integral and the volume integral give the answer 1 so the divergence theorem is verified.

5.3 In order to use the divergence theorem, the volume integral must first be written in terms of a divergence. This can be done using (4.27):

$$\iiint_V \boldsymbol{u} \cdot \nabla\phi\, dV = \iiint \nabla \cdot (\phi \boldsymbol{u}) - \phi \nabla \cdot \boldsymbol{u}\, dV.$$

Now since the fluid is incompressible, $\nabla \cdot \boldsymbol{u} = 0$. Applying the divergence theorem then gives

$$\iiint_V \boldsymbol{u} \cdot \nabla\phi\, dV = \oiint_S \phi \boldsymbol{u} \cdot \boldsymbol{n}\, dS.$$

Since it is given that $\boldsymbol{u} \cdot \boldsymbol{n} = 0$ on S, the value of the integral is zero.

5.4 This result follows directly from applying the divergence theorem to the vector field ∇f, since $\nabla \cdot \nabla f = \nabla^2 f = g$.

5.5 To apply the divergence theorem, a closed surface must be used. Let S' be the surface $z = 0$, $x^2 + y^2 < 1$, forming the base of the hemisphere. The divergence theorem can now be applied over the entire closed surface

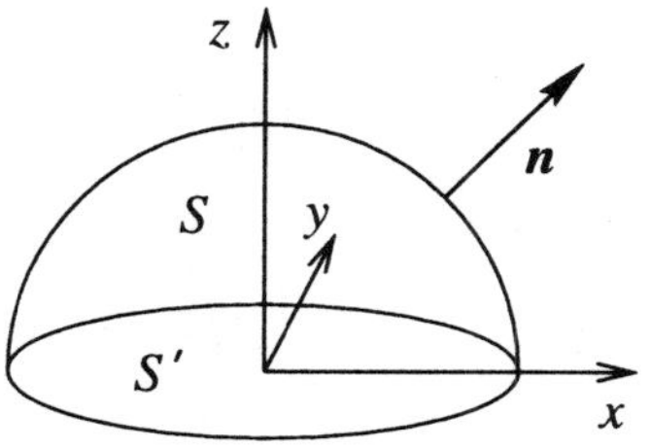

$S + S'$, giving

$$\iiint_V \nabla \cdot \boldsymbol{v}\, dV = \iint_S \boldsymbol{v} \cdot \boldsymbol{n}\, dS + \iint_{S'} \boldsymbol{v} \cdot \boldsymbol{n}\, dS.$$

The surface integral over S can therefore be found by subtracting the surface integral over S' from the volume integral.
For $\boldsymbol{v} = (x + y, z^2, x^2)$, $\nabla \cdot \boldsymbol{v} = 1$, so the volume integral is just the volume of the hemisphere, $2\pi/3$. The surface integral over S' is

$$\iint_{S'} \boldsymbol{v} \cdot \boldsymbol{n}\, dS = \iint_{S'} -x^2\, dS$$

since $\boldsymbol{n} = (0, 0, -1)$ on S'. This integral can be evaluated using polar coordinates (r, θ), where $x = r\cos\theta$, $0 \leq r \leq 1$, $0 \leq \theta \leq 2\pi$ and $dS = r\, d\theta\, dr$ (see Section 2.3.2).

$$\begin{aligned} \iint_{S'} -x^2\, dS &= \int_0^1 \int_0^{2\pi} -r^2 \cos^2\theta\, r\, d\theta\, dr \\ &= \int_0^1 -r^3 \pi\, dr \\ &= -\pi/4. \end{aligned}$$

The required integral is then

$$\iint_S \boldsymbol{v} \cdot \boldsymbol{n}\, dS = 2\pi/3 + \pi/4 = 11\pi/12.$$

5.6 The argument follows that of Section 5.1.1 with ρ replaced by q and $\rho\boldsymbol{u}$ replaced by $\boldsymbol{j}$, so the law of conservation of electric charge is

$$\frac{\partial q}{\partial t} + \boldsymbol{\nabla} \cdot \boldsymbol{j} = 0.$$

5.7 Consider (5.13) in the case where the volume δV is small. The volume integral of $\boldsymbol{\nabla} f$ is then approximately equal to $\boldsymbol{\nabla} f$ times the volume:

$$\boldsymbol{\nabla} f \, \delta V \approx \oiint_{\delta S} f \boldsymbol{n} \, dS.$$

Dividing by the volume and taking the limit $\delta V \to 0$ gives the definition

$$\boldsymbol{\nabla} f = \lim_{\delta V \to 0} \frac{1}{\delta V} \oiint_{\delta S} f \boldsymbol{n} \, dS,$$

which is analogous in form to the original definitions of div and curl.

5.8 Using Stokes's theorem,

$$\oint_C \boldsymbol{r} \cdot d\boldsymbol{r} = \iint_S \boldsymbol{\nabla} \times \boldsymbol{r} \cdot \boldsymbol{n} \, dS = 0$$

since $\boldsymbol{\nabla} \times \boldsymbol{r} = \boldsymbol{0}$.

5.9 First consider the line integral around the circle $x^2 + y^2 = 1$, evaluated parametrically using $x = \cos\theta$, $y = \sin\theta$, $0 \le \theta \le 2\pi$, with $d\boldsymbol{r} = (-\sin\theta, \cos\theta, 0)$. The value of the line integral is

$$\begin{aligned}
&\int_0^{2\pi} (2\cos\theta - \sin\theta, \; -\sin^2\theta, 0) \cdot (-\sin\theta, \cos\theta, 0) \, d\theta \\
&\quad = \int_0^{2\pi} -2\cos\theta\sin\theta + \sin^2\theta - \sin^2\theta\cos\theta \, d\theta \\
&\quad = \int_0^{2\pi} -\sin 2\theta + (1 - \cos 2\theta)/2 - \sin^2\theta\cos\theta \, d\theta \\
&\quad = \pi,
\end{aligned}$$

since all terms except the $1/2$ give zero when integrated between 0 and 2π. Now to compute the surface integral we need $(\boldsymbol{\nabla} \times \boldsymbol{u}) \cdot \boldsymbol{n}$. The line integral was taken in an anticlockwise sense in the x,y plane, so the right-hand rule means that $\boldsymbol{n}$ points in the positive z direction, so $(\boldsymbol{\nabla} \times \boldsymbol{u}) \cdot \boldsymbol{n} = (\boldsymbol{\nabla} \times \boldsymbol{u}) \cdot \boldsymbol{e}_3 = 1$. The value of the surface integral is then just the area of the surface, which is π since the surface is a disk of radius 1. Thus Stokes's theorem is verified since the surface integral and the line integral are both equal to π.

5.10 Applying Stokes's theorem,

$$\begin{aligned}\oint_C f\boldsymbol{\nabla} g\cdot d\boldsymbol{r} &= \iint_S \boldsymbol{\nabla}\times(f\boldsymbol{\nabla} g)\cdot\boldsymbol{n}\,dS\\ &= \iint_S(\boldsymbol{\nabla} f\times\boldsymbol{\nabla} g+f\boldsymbol{\nabla}\times(\boldsymbol{\nabla} g))\cdot\boldsymbol{n}\,dS\\ &= \iint_S(\boldsymbol{\nabla} f\times\boldsymbol{\nabla} g)\cdot\boldsymbol{n}\,dS.\end{aligned}$$

Similarly,

$$\oint_C g\boldsymbol{\nabla} f\cdot d\boldsymbol{r}=\iint_S(\boldsymbol{\nabla} g\times\boldsymbol{\nabla} f)\cdot\boldsymbol{n}\,dS=-\iint_S(\boldsymbol{\nabla} f\times\boldsymbol{\nabla} g)\cdot\boldsymbol{n}\,dS.$$

Therefore

$$\oint_C f\boldsymbol{\nabla} g\cdot d\boldsymbol{r}=-\oint_C g\boldsymbol{\nabla} f\cdot d\boldsymbol{r}.$$

Alternatively, this result can be obtained by applying Stokes's theorem to the line integral of $\boldsymbol{\nabla}(fg)$.

5.11 If $\boldsymbol{u}$ is irrotational, $\boldsymbol{\nabla}\times\boldsymbol{u}=\boldsymbol{0}$ and so $\boldsymbol{u}\times\boldsymbol{\nabla} f=-\boldsymbol{\nabla}\times(f\boldsymbol{u})$. Applying Stokes's theorem, the surface integral is equal to $\oint_C -f\boldsymbol{u}\cdot d\boldsymbol{r}$.

5.12 The rate of change of the total magnetic flux through a surface S is

$$\frac{d}{dt}\iint_S \boldsymbol{B}\cdot\boldsymbol{n}\,dS=\iint_S \boldsymbol{\nabla}\times(\boldsymbol{u}\times\boldsymbol{B})\cdot\boldsymbol{n}\,dS=\oint_C \boldsymbol{u}\times\boldsymbol{B}\cdot d\boldsymbol{r}.$$

The curve C is a streamline so $\boldsymbol{u}$ is parallel to $d\boldsymbol{r}$ and hence the scalar triple product that appears in the line integral is zero. Therefore the flux of $\boldsymbol{B}$ through S does not change with time.

5.13 Applying (5.18) with $\boldsymbol{v}=\boldsymbol{r}$ gives

$$\iint_S \frac{\partial x_k}{\partial x_k}n_j-\frac{\partial x_k}{\partial x_j}n_k\,dS=\left[\oint_C \boldsymbol{v}\times d\boldsymbol{r}\right]_j.$$

Using $\partial x_k/\partial x_j=\delta_{jk}$ and $\delta_{jj}=3$, this simplifies to

$$\iint_S 2n_j\,dS=\left[\oint_C \boldsymbol{v}\times d\boldsymbol{r}\right]_j.$$

For a flat surface n_j is constant so the l.h.s. has the value $2An_j$, giving the required result.

Solutions to Exercises for Chapter 6

6.1 For Cartesian coordinates, $u_i = x_i$. The scale factor h_1 is, from (6.2),

$$h_1 = \left|\frac{\partial \boldsymbol{x}}{\partial x_1}\right| = \left|(\frac{\partial x_1}{\partial x_1}, \frac{\partial x_2}{\partial x_1}, \frac{\partial x_3}{\partial x_1})\right| = |(1,0,0)| = 1.$$

By the same reasoning h_2 and h_3 are also 1.

6.2 (a) Using the definition of the scale factors,

$$\begin{aligned} h_u &= \left|(vw, v(1-w^2)^{1/2}, u)\right| = (u^2+v^2)^{1/2}, \\ h_v &= \left|(uw, u(1-w^2)^{1/2}, -v)\right| = (u^2+v^2)^{1/2}, \\ h_w &= \left|(uv, -uvw(1-w^2)^{-1/2}, 0)\right| = uv/(1-w^2)^{1/2}. \end{aligned}$$

(b) To show that the (u, v, w) system is orthogonal we first need to compute the unit vectors. From (6.1) and the above results for the scale factors, these are

$$\begin{aligned} \boldsymbol{e_u} &= (vw, v(1-w^2)^{1/2}, u)/(u^2+v^2)^{1/2}, \\ \boldsymbol{e_v} &= (uw, u(1-w^2)^{1/2}, -v)/(u^2+v^2)^{1/2}, \\ \boldsymbol{e_w} &= ((1-w^2)^{1/2}, -w, 0). \end{aligned}$$

Now take the dot product of these unit vectors to check that they are orthogonal:

$$\boldsymbol{e_u} \cdot \boldsymbol{e_v} = (uvw^2 + uv(1-w^2) - uv)/(u^2+v^2) = 0,$$

and similarly $\boldsymbol{e_u} \cdot \boldsymbol{e_w} = 0$ and $\boldsymbol{e_v} \cdot \boldsymbol{e_w} = 0$.

(c) The volume element in the (u, v, w) system is

$$dV = h_1 h_2 h_3 \, du \, dv \, dw = (u^2+v^2)uv/(1-w^2)^{1/2} \, du \, dv \, dw.$$

6.3 Following the approach of the previous exercise, the scale factors are found to be $h_u = (u^2+v^2)^{1/2}$, $h_v = (u^2+v^2)^{1/2}$ and $h_\theta = uv$, so the volume element is $dV = (u^2+v^2)uv$. The volume V between the surfaces $u = 1$ and $v = 1$ is therefore

$$V = \int_0^1 \int_0^1 \int_0^{2\pi} (u^2+v^2)uv \, d\theta \, du \, dv = 2\pi \int_0^1 \left[u^4 v/4 + u^2 v^3/2\right]_0^1 dv$$

which yields the result $V = \pi/2$.

6.4 In Cartesian coordinates, $\boldsymbol{\nabla} f$ and $\boldsymbol{e_1}$ are

$$\boldsymbol{\nabla} f = \left(\frac{\partial f}{\partial x_1}, \frac{\partial f}{\partial x_2}, \frac{\partial f}{\partial x_3}\right), \quad \boldsymbol{e_1} = \frac{\partial \boldsymbol{x}}{\partial u_1}\Big/ h_1 = \frac{1}{h_1}\left(\frac{\partial x_1}{\partial u_1}, \frac{\partial x_2}{\partial u_1}, \frac{\partial x_3}{\partial u_1}\right)$$

so the component of $\boldsymbol{\nabla} f$ in the $\boldsymbol{e_1}$ direction is

$$\boldsymbol{e_1} \cdot \boldsymbol{\nabla} f = \frac{1}{h_1}\left(\frac{\partial f}{\partial x_1}\frac{\partial x_1}{\partial u_1} + \frac{\partial f}{\partial x_2}\frac{\partial x_2}{\partial u_1} + \frac{\partial f}{\partial x_3}\frac{\partial x_3}{\partial u_1}\right) = \frac{1}{h_1}\frac{\partial f}{\partial u_1}.$$

Similar expressions for the $\boldsymbol{e_2}$ and $\boldsymbol{e_3}$ components follow, giving the formula (6.8) for $\boldsymbol{\nabla} f$ in an orthogonal curvilinear coordinate system.

6.5 Assume that the cylinder is aligned with the centre of the apple and that $a < b$. The volume integral can be carried out using either cylindrical or spherical coordinates. Using cylindrical coordinates, the limits on the coordinates R and ϕ are $0 < R < a$, $0 < \phi < 2\pi$. The limits on z are determined by the radius of the sphere, $z^2 + R^2 = b^2$, so the limits are $-\sqrt{b^2 - R^2} < z < \sqrt{b^2 - R^2}$. The required volume is

$$\begin{aligned}
\int_0^a \int_{-\sqrt{b^2-R^2}}^{\sqrt{b^2-R^2}} \int_0^{2\pi} R\, d\phi\, dz\, dR &= 2\pi \int_0^a 2R\sqrt{b^2 - R^2}\, dR \\
&= 4\pi\big[-(b^2 - R^2)^{3/2}/3\big]_0^a \\
&= 4\pi(b^3 - (b^2 - a^2)^{3/2})/3.
\end{aligned}$$

The proportion of the apple removed is this volume divided by $4\pi b^3/3$, which is $1 - (1 - a^2/b^2)^{3/2}$.

6.6 The limits on the coordinates are $\pi/3 \le \theta \le 2\pi/3$, $0 \le \phi < 2\pi$, $0 \le r \le a$, where a is the radius of the Earth. The volume V is

$$V = \int_{\pi/3}^{2\pi/3} \int_0^{2\pi} \int_0^a r^2 \sin\theta\, dr\, d\phi\, d\theta = 2\pi\left[r^3/3\right]_0^a \left[-\cos\theta\right]_{\pi/3}^{2\pi/3} = 2\pi a^3/3,$$

so exactly half of the Earth's volume is less than 30° away from the Equator.

6.7 In spherical polar coordinates the divergence of $\boldsymbol{e_\phi}$ is zero, using (6.23). The curl, from (6.25), is

$$\boldsymbol{\nabla} \times \boldsymbol{e_\phi} = \frac{1}{r\sin\theta}\frac{\partial \sin\theta}{\partial\theta}\boldsymbol{e_r} - \frac{1}{r}\frac{\partial r}{\partial r}\boldsymbol{e_\theta} = \frac{\cot\theta}{r}\boldsymbol{e_r} - \frac{1}{r}\boldsymbol{e_\theta}.$$

6.8 The formula $\boldsymbol{u} \cdot \boldsymbol{\nabla u} = \boldsymbol{\nabla}(|\boldsymbol{u}|^2/2) - \boldsymbol{u} \times (\boldsymbol{\nabla} \times \boldsymbol{u})$ is used to find $\boldsymbol{u} \cdot \boldsymbol{\nabla u}$. Since $\boldsymbol{u}$ is a unit vector, its magnitude is constant so $\boldsymbol{\nabla}(|\boldsymbol{u}|^2/2) = \boldsymbol{0}$. Using (6.17), $\boldsymbol{\nabla} \times \boldsymbol{u} = \boldsymbol{e_z}/R$, so $\boldsymbol{u} \times (\boldsymbol{\nabla} \times \boldsymbol{u}) = (0, 1, 0) \times (0, 0, 1/R) = (1/R, 0, 0)$. Therefore $\boldsymbol{u} \cdot \boldsymbol{\nabla u} = -\boldsymbol{e_R}/R$.

6.9 Recall that the definition of the Laplacian of a vector field is $\nabla^2 \boldsymbol{v} = \boldsymbol{\nabla\nabla} \cdot \boldsymbol{v} - \boldsymbol{\nabla} \times \boldsymbol{\nabla} \times \boldsymbol{v}$. Using the formulae (6.14) and (6.15), the R component of $\boldsymbol{\nabla\nabla} \cdot \boldsymbol{v}$ is

$$\frac{\partial}{\partial R}\left(\frac{v_R}{R} + \frac{\partial v_R}{\partial R} + \frac{1}{R}\frac{\partial v_\phi}{\partial \phi} + \frac{\partial v_z}{\partial z}\right)$$

which can be expanded to give

$$\frac{1}{R}\frac{\partial v_R}{\partial R} - \frac{v_R}{R}^2 + \frac{\partial^2 v_R}{\partial R^2} - \frac{1}{R^2}\frac{\partial v_\phi}{\partial \phi} + \frac{1}{R}\frac{\partial^2 v_\phi}{\partial \phi \partial R} + \frac{\partial^2 v_z}{\partial R \partial z}.$$

Computing the R component of $\boldsymbol{\nabla} \times \boldsymbol{\nabla} \times \boldsymbol{v}$ using (6.17) gives

$$\frac{1}{R^2}\frac{\partial}{\partial \phi}\left(v_\phi + R\frac{\partial v_\phi}{\partial R}\right) - \frac{1}{R^2}\frac{\partial^2 v_R}{\partial \phi^2} - \frac{\partial^2 v_R}{\partial z^2} + \frac{\partial^2 v_z}{\partial R \partial z}.$$

Subtracting these two quantities gives the R component of $\nabla^2 \boldsymbol{v}$:

$$(\nabla^2 \boldsymbol{v})_R = \frac{1}{R}\frac{\partial v_R}{\partial R} - \frac{v_R}{R^2} + \frac{\partial^2 v_R}{\partial R^2} - \frac{2}{R^2}\frac{\partial v_\phi}{\partial \phi} + \frac{1}{R^2}\frac{\partial^2 v_R}{\partial \phi^2} + \frac{\partial^2 v_R}{\partial z^2}.$$

Note that this cannot be the Laplacian of the R component of $\boldsymbol{v}$, since it involves v_ϕ. In fact $(\nabla^2 \boldsymbol{v})_R$ and $\nabla^2(v_R)$ are related by

$$(\nabla^2 \boldsymbol{v})_R = \nabla^2(v_R) - \frac{v_R}{R^2} - \frac{2}{R^2}\frac{\partial v_\phi}{\partial \phi}.$$

Solutions to Exercises for Chapter 7

7.1 The definition $L_{ij} = \boldsymbol{e}'_i \cdot \boldsymbol{e}_j$ states that L_{ij} is the cosine of the angle between the $\boldsymbol{e}'_i$ and $\boldsymbol{e}_j$ unit vectors. Referring to Figure 7.1, the cosine of the angle between $\boldsymbol{e}'_1$ and $\boldsymbol{e}_1$ is $\cos\theta$, and the cosine of the angle between $\boldsymbol{e}'_2$ and $\boldsymbol{e}_2$ is the same. The cosine of the angle between $\boldsymbol{e}'_1$ and $\boldsymbol{e}_2$ is $\cos(\pi/2-\theta) = \sin\theta$, and the cosine of the angle between $\boldsymbol{e}'_2$ and $\boldsymbol{e}_1$ is $\cos(\pi/2+\theta) = -\sin\theta$. These results agree with the matrix in (7.3).

7.2 $\boldsymbol{u}$ is a vector, so $u'_i = L_{ij}u_j$. To show $\boldsymbol{\nabla} \cdot \boldsymbol{u}$ is a scalar we need to compute its value in the dashed frame:

$$(\boldsymbol{\nabla} \cdot \boldsymbol{u})' = \frac{\partial u'_i}{\partial x'_i} = \frac{\partial}{\partial x'_i}(L_{ij}u_j) = L_{ij}\frac{\partial u_j}{\partial x_k}\frac{\partial x_k}{\partial x'_i},$$

using the chain rule. Now using (7.8),

$$(\boldsymbol{\nabla} \cdot \boldsymbol{u})' = L_{ij}\frac{\partial u_j}{\partial x_k}L_{ik} = \delta_{jk}\frac{\partial u_j}{\partial x_k} = \frac{\partial u_j}{\partial x_j} = \boldsymbol{\nabla} \cdot \boldsymbol{u}.$$

Thus $\nabla \cdot \boldsymbol{u}$ has the same value in the dashed and undashed frame, so $\nabla \cdot \boldsymbol{u}$ is a scalar.

7.3 Since $\boldsymbol{a}$ and $\boldsymbol{b}$ are vectors, $a_i' = L_{ik} a_k$ and $b_j' = L_{jm} b_m$. Note that the suffices are carefully chosen to avoid repetition. Then

$$(a_i b_j)' = a_i' b_j' = L_{ik} L_{jm} a_k b_m.$$

This agrees with the transformation rule (7.13), so $a_i b_j$ is a second-rank tensor.

7.4 Note that T_{ij} can be written as $T_{ij} = a_i b_j$ where a is the position vector (x_1, x_2) and $b = (x_2, -x_1)$ which was shown to be a vector in Example 7.3. Therefore T_{ij} is a tensor by the result of the previous question. Alternatively, the fact that T_{ij} is a tensor can be confirmed by following the method of Example 7.3.

7.5 Since ϕ is a scalar, $\phi' = \phi$. The transformation rule for T_{jk} is

$$T_{jk}' = \frac{\partial^2 \phi'}{\partial x_j' \partial x_k'} = \frac{\partial^2 \phi}{\partial x_m \partial x_n} \frac{\partial x_m}{\partial x_j'} \frac{\partial x_n}{\partial x_k'} = \frac{\partial^2 \phi}{\partial x_m \partial x_n} L_{jm} L_{kn},$$

so T_{jk} is a second-rank tensor.

7.6 If T_{ij} is a tensor, T_{ij} obeys the rule

$$T_{ij}' = L_{ik} L_{jm} T_{km}.$$

Setting $j = i$ in this formula,

$$T_{ii}' = L_{ik} L_{im} T_{km} = \delta_{km} T_{km} = T_{kk} = T_{ii}.$$

So the value of T_{ii} is the same in the dashed and undashed frames, i.e. T_{ii} is a scalar.

7.7 In suffix notation, the divergence theorem (5.1) becomes

$$\iiint_V \frac{\partial u_j}{\partial x_j} \, dV = \oiint_S u_j n_j \, dS.$$

This result holds if u_j is replaced by T_{1j}, T_{2j} or T_{3j}, giving the required result (7.16).

7.8 If Q_{ijkl} is a tensor of rank four,

$$Q_{ijkl}' = L_{ip} L_{jq} L_{kr} L_{ls} Q_{pqrs}.$$

7.9 Q_{ijkl} obeys the above rule, so by setting $k = j$,

$$Q_{ijjl}' = L_{ip} L_{jq} L_{jr} L_{ls} Q_{pqrs} = L_{ip} \delta_{qr} L_{ls} Q_{pqrs} = L_{ip} L_{ls} Q_{prrs},$$

so Q_{ijjl} obeys the transformation law for a tensor of rank two.

7.10 Given that $u_i a_i$ is a scalar,

$$u'_i a'_i = u_j a_j = u_j L_{ij} a'_i$$

so $(u'_i - u_j L_{ij})a'_i = 0$. If this holds for any a_i, then $u'_i = L_{ij}u_j$, so u_i is a vector.

7.11 B_{rs} is an anti-symmetric tensor, so

$$B'_{rs} = L_{ru}L_{sv}B_{uv} = -L_{sv}L_{ru}B_{vu} = -B'_{sr}.$$

Hence B is also anti-symmetric in the dashed frame.

7.12 If B_{rs} is anti-symmetric, $B_{rs} = -B_{sr}$ so $B_{rr} = -B_{rr}$ and hence $B_{rr} = 0$.

7.13 A_{ijk} has the properties $A_{ijk} = A_{jik}$ and $A_{ijk} = -A_{ikj}$. Repeatedly applying these rules alternately gives

$$A_{ijk} = A_{jik} = -A_{jki} = -A_{kji} = A_{kij} = A_{ikj} = -A_{ijk}.$$

So any element of A_{ijk} is equal to minus itself, hence all elements are zero.

7.14 (a) $A_{ij} = \epsilon_{ijk}B_k$, so

$$\begin{aligned} A'_{ij} &= \epsilon'_{ijk}B'_k \\ &= L_{im}L_{jn}L_{kp}\epsilon_{mnp}L_{kr}B_r \\ &= L_{im}L_{jn}\delta_{pr}\epsilon_{mnp}B_r \\ &= L_{im}L_{jn}\epsilon_{mnp}B_p = L_{im}L_{jn}A_{mn}, \end{aligned}$$

so A_{ij} obeys the transformation rule for a second-rank tensor. Since ϵ_{ijk} is anti-symmetric with respect to any two indices, A_{ij} is antisymmetric.

(b) Multiply through by ϵ_{ijm}:

$$\epsilon_{ijm}A_{ij} = \epsilon_{ijm}\epsilon_{ijk}B_k = 2\delta_{km}B_k = 2B_m,$$

making use of Exercise 4.4(c). Hence $B_m = \epsilon_{ijm}A_{ij}/2$.

7.15 The most general isotropic fourth-rank tensor is $a_{ijkl} = \lambda\delta_{ij}\delta_{kl} + \mu\delta_{ik}\delta_{jl} + \nu\delta_{il}\delta_{jk}$, from Theorem 7.4. From (4.12), the difference between products of δ_{ij} terms can be written in terms of a products of ϵ_{ijk} terms. For example, if $\lambda = 1$, $\mu = -1$, $\nu = 0$, we have the isotropic tensor $a_{ijkl} = \delta_{ij}\delta_{kl} - \delta_{ik}\delta_{jl} = \epsilon_{ilm}\epsilon_{mjk}$.

7.16 Since δ_{ij} and ϵ_{ijk} are isotropic tensors, the combination $a_{ijklm} = \delta_{ij}\epsilon_{klm}$ is an isotropic fifth-rank tensor. Since any two of the five suffices can be chosen for the δ, for example $a_{ijklm} = \delta_{ik}\epsilon_{jlm}$ or $a_{ijklm} = \delta_{il}\epsilon_{jkm}$, the total number of different tensors of this type is the number of ways of choosing two objects from five, which is $5!/2!(5-2)! = 10$. So there are at least ten different components in the most general isotropic fifth-rank tensor.

7.17 Following the approach of Section 7.4.2, the kinetic energy of a volume element dV is $\rho\,|\boldsymbol{v}|^2 dV/2$ and $\boldsymbol{v} = \boldsymbol{\Omega} \times \boldsymbol{r}$, so

$$\begin{aligned}
E &= 1/2 \iiint_V \rho\,|\boldsymbol{v}|^2\, dV \\
&= 1/2 \iiint_V \rho\,\epsilon_{ijk}\Omega_j r_k \epsilon_{ilm}\Omega_l r_m\, dV \\
&= 1/2 \iiint_V \rho\,(\Omega_j\Omega_j r_k r_k - \Omega_j r_j \Omega_k r_k)\, dV \\
&= 1/2 \iiint_V \rho\,(\delta_{jk} r^2 - r_j r_k)\,\Omega_j\Omega_k\, dV \\
&= I_{jk}\Omega_j\Omega_k/2.
\end{aligned}$$

Solutions to Exercises for Chapter 8

8.1 The size of the body L is a length and the diffusivity k has units of length^2/time. The only combination of these which has the units of time is L^2/k. Therefore the time for heat to diffuse through a body is proportional to the square of the size of the body.
(a) Assume that the mammoth is the same shape as the chicken, made of the same material (so the wooliness is ignored) and 20 times the length of the chicken. The defrosting time is therefore $20^2 \times 6$ hours which is 100 days.
(b) Assume that to cook properly, a certain temperature must be reached in the interior. The time required for this is proportional to L^2. The mass M is proportional to L^3, so $L \propto M^{1/3}$. Therefore the cooking time should be proportional to the two-thirds power of the mass: $t \propto M^{2/3}$.

8.2 Using (8.4),

$$\begin{aligned}
\frac{\partial\rho}{\partial t} &= \epsilon_0 \boldsymbol{\nabla}\cdot\frac{\partial \boldsymbol{E}}{\partial t} \\
&= \epsilon_0 \boldsymbol{\nabla}\cdot(\boldsymbol{\nabla}\times\boldsymbol{B} - \mu_0\boldsymbol{j})/\epsilon_0\mu_0 \quad \text{from (8.7)} \\
&= -\boldsymbol{\nabla}\cdot\boldsymbol{j}
\end{aligned}$$

since the combination div curl is always zero.

8.3 Gauss's law says that the total flux of electric field through the surface is the total charge within the surface divided by ϵ_0. Taking the surface to be the surface of a sphere of radius r, with area $4\pi r^2$, Gauss's law gives $4\pi r^2 E_r = Q/\epsilon_0$, which agrees with the result of Example 8.4.

8.4 Take the curl of (8.12):

$$\nabla \times (\nabla \times \boldsymbol{B}) = \mu_0 \epsilon_0 \frac{\partial \nabla \times \boldsymbol{E}}{\partial t}.$$

Expanding the r.h.s. and using (8.10) and (8.11) gives

$$-\nabla^2 \boldsymbol{B} = -\mu_0 \epsilon_0 \frac{\partial^2 \boldsymbol{B}}{\partial t^2},$$

so $\boldsymbol{B}$ obeys exactly the same wave equation as $\boldsymbol{E}$.

8.5 Given the electric field $\boldsymbol{E} = \boldsymbol{E}_0 f(\boldsymbol{k}\cdot\boldsymbol{x} - \omega t) = \boldsymbol{E}_0 f(u)$ where $u = \boldsymbol{k}\cdot\boldsymbol{x} - \omega t$, the magnetic field $\boldsymbol{B}$ can be found using (8.11):

$$\begin{aligned} \nabla \times \boldsymbol{E} &= \nabla \times (\boldsymbol{E}_0 f(u)) \\ &= \nabla f \times \boldsymbol{E}_0 \\ &= \frac{df}{du} \boldsymbol{k} \times \boldsymbol{E}_0. \end{aligned}$$

Integrating with respect to t and changing the sign to find $\boldsymbol{B}$ gives

$$\boldsymbol{B} = \frac{1}{\omega} \boldsymbol{k} \times \boldsymbol{E}_0 f(\boldsymbol{k}\cdot\boldsymbol{x} - \omega t) = \frac{1}{\omega} \boldsymbol{k} \times \boldsymbol{E}.$$

Hence the magnetic field $\boldsymbol{B}$ is perpendicular to the electric field $\boldsymbol{E}$.

8.6 The energy can be written $w = \boldsymbol{B}\cdot\boldsymbol{B}/2 + \boldsymbol{E}\cdot\boldsymbol{E}/2c^2$, so the rate of change of energy is

$$\begin{aligned} \frac{\partial w}{\partial t} &= \boldsymbol{B}\cdot\frac{\partial \boldsymbol{B}}{\partial t} + \frac{1}{c^2}\boldsymbol{E}\cdot\frac{\partial \boldsymbol{E}}{\partial t} \\ &= -\boldsymbol{B}\cdot\nabla\times\boldsymbol{E} + \boldsymbol{E}\cdot\nabla\times\boldsymbol{B} \quad \text{using (8.11) and (8.12)} \\ &= -\nabla\cdot(\boldsymbol{E}\times\boldsymbol{B}). \end{aligned}$$

This gives the conservation law

$$\frac{\partial w}{\partial t} + \nabla\cdot\boldsymbol{P} = 0$$

where $\boldsymbol{P} = \boldsymbol{E}\times\boldsymbol{B}$ is known as the Poynting vector, representing the energy flux of the electromagnetic wave.

8.7 (a) If the material is in equilibrium then the stress tensor must be symmetric (8.16), so $a = 0$ and $c = b$. Also, $\partial P_{ij}/\partial x_j = 0$. For the first row ($i = 1$) this is satisfied. The second row ($i = 2$) gives $m x_2^{m-1} + b x_2 = 0$, so $m = 2$ and $b = -2$. From the third row, $c x_3 + n x_3^{n-1} = 0$, so $n = 2$ and $c = -2$.
(b) For the surface $x_1 = 1$ the normal is $(1, 0, 0)$ (the normal points in the direction of the side which is exerting the force). Thus the force is $F_i = P_{ij} n_j \, dS = P_{i1} \, dS$. Since $P_{11} = P_{21} = 0$ and $P_{31} = x_2$, the force is only in the x_3 direction. Its magnitude is

$$F_3 = \int_0^1 \int_0^1 x_2 \, dx_2 \, dx_3 = 1/2.$$

8.8 The r.h.s. of (8.22) is

$$\begin{aligned}
-\frac{1}{2}\epsilon_{ijk}\left[\boldsymbol{\nabla}\times\boldsymbol{v}\right]_k &= -\frac{1}{2}\epsilon_{ijk}\epsilon_{klm}\frac{\partial v_m}{\partial x_l} \\
&= -\frac{1}{2}(\delta_{il}\delta_{jm}-\delta_{im}\delta_{jl})\frac{\partial v_m}{\partial x_l} \\
&= \frac{1}{2}\left(\frac{\partial v_i}{\partial x_j}-\frac{\partial v_j}{\partial x_i}\right) = S_{ij}.
\end{aligned}$$

8.9 Applying the condition $\partial P_{ij}/\partial x_j = 0$ and using (8.21),

$$\begin{aligned}
0 &= \lambda\delta_{ij}\frac{\partial E_{kk}}{\partial x_j} + 2\mu\frac{\partial E_{ij}}{\partial x_j} \\
&= \lambda\frac{\partial}{\partial x_i}\frac{\partial v_j}{\partial x_j} + \mu\left(\frac{\partial^2 v_i}{\partial x_j \partial x_j} + \frac{\partial^2 v_j}{\partial x_i \partial x_j}\right) \\
&= (\lambda+\mu)\frac{\partial}{\partial x_i}\boldsymbol{\nabla}\cdot\boldsymbol{v} + \mu\nabla^2 v_i.
\end{aligned}$$

8.10 (a) From the definition of the strain tensor, $E_{11} = ax_2$, $E_{12} = E_{21} = ax_1/2 + bx_1$, $E_{22} = -2bx_2$ and the other components of E_{ij} are zero.
(b) Using (8.21), $P_{11} = \lambda(a-2b)x_2 + 2\mu a x_2$, $P_{12} = P_{21} = 2\mu(ax_1/2 + bx_1)$, $P_{22} = \lambda(a-2b)x_2 - 4\mu b x_2$, $P_{33} = \lambda(a-2b)x_2$ and other components of P_{ij} are zero.
(c) Applying the equilibrium condition $\partial P_{ij}/\partial x_j = 0$, this is identically satisfied for $i=1$ and $i=3$, but for $i=2$, $\mu a + 2\mu b + \lambda(a-2b) - 4\mu b = 0$. This is satisfied if $a = 2b$.

8.11 (a) The term $\boldsymbol{\nabla}\cdot(\rho\boldsymbol{u})$ in (5.9) can be written $\boldsymbol{u}\cdot\boldsymbol{\nabla}(\rho_0+\rho_1)+(\rho_0+\rho_1)\boldsymbol{\nabla}\cdot\boldsymbol{u}$. Now since ρ_0 is a constant and terms involving products of $\boldsymbol{u}$ and ρ_1 can be neglected, this simplifies to $\rho_0\boldsymbol{\nabla}\cdot\boldsymbol{u}$, so (5.9) becomes $\partial\rho_1/\partial t + \rho_0\boldsymbol{\nabla}\cdot\boldsymbol{u} = 0$.
(b) In the Navier–Stokes equation (8.28) the product term $\boldsymbol{u}\cdot\boldsymbol{\nabla}\boldsymbol{u}$ can be ignored and there is no body force $\boldsymbol{b}$ or viscosity μ, so the remaining terms are $(\rho_0+\rho_1)(\partial\boldsymbol{u}/\partial t) = -\boldsymbol{\nabla}(p_0+p_1)$. The term involving ρ_1 and $\boldsymbol{u}$ can be ignored, p_0 is constant and $p_1 = a\rho_1$, so $\rho_0\partial\boldsymbol{u}/\partial t = -a\boldsymbol{\nabla}\rho_1$.
(c) The velocity $\boldsymbol{u}$ can be eliminated from (8.35) and (8.36) by taking the time derivative of (8.35):

$$\frac{\partial^2\rho_1}{\partial t^2} = -\rho_0\boldsymbol{\nabla}\cdot\frac{\partial\boldsymbol{u}}{\partial t} = a\boldsymbol{\nabla}\cdot\boldsymbol{\nabla}\rho_1 = a\nabla^2\rho_1.$$

Hence the density perturbation obeys the wave equation (8.13). Physically, the waves are sound waves travelling through the fluid.

Index